IMAGES
of America

THE DAMS OF WESTERN SAN DIEGO COUNTY

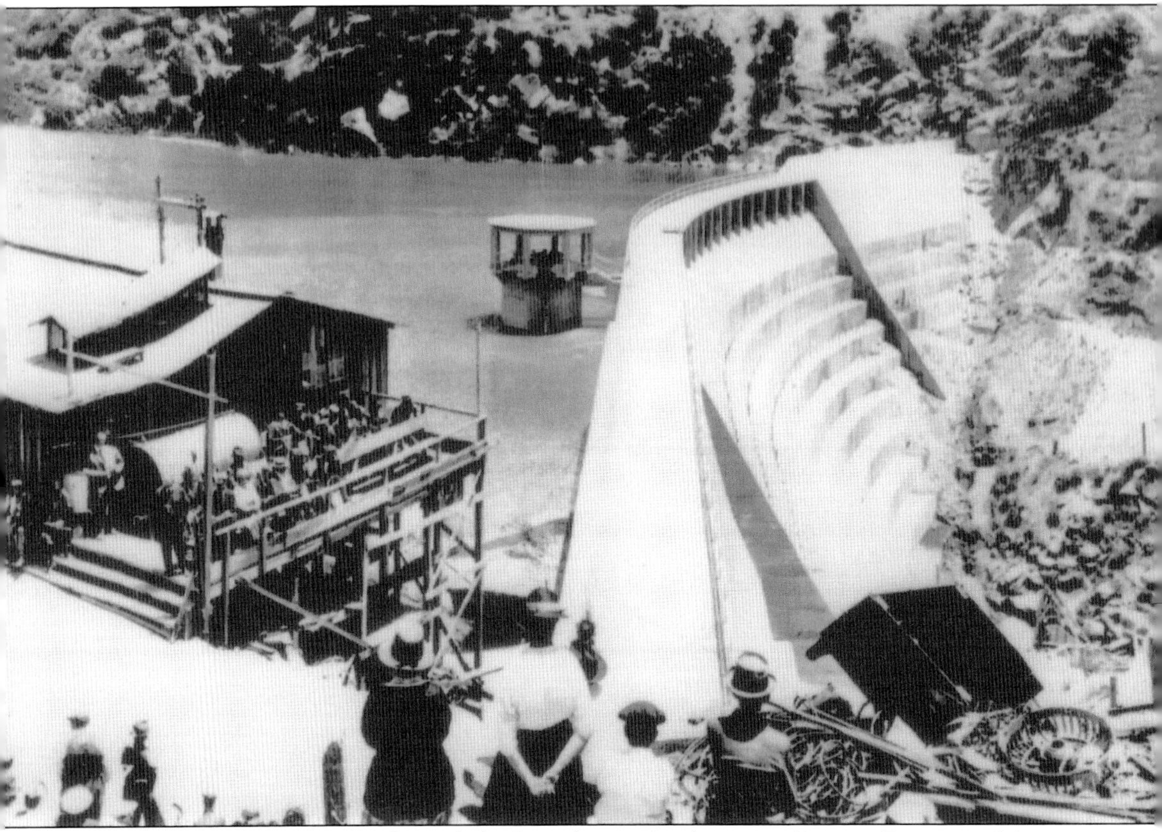

Hundreds of San Diegans braved the hot, dusty 35-mile trip to Barrett Dam to celebrate its dedication on July 25, 1922. The heavy spring rains that had interfered with construction had now almost filled the reservoir. (Courtesy of San Diego City Library, California Room Collection.)

On the Cover: A foreman and a crew of workers pose on temporary steps leading down to the Loveland Dam construction site in 1945. The completed lower portion of the concrete dam has started to conserve the Sweetwater River and form the long, narrow Loveland Reservoir. (Courtesy of Bonita Museum and Cultural Center.)

IMAGES
of America

THE DAMS OF WESTERN SAN DIEGO COUNTY

John Martin

ARCADIA
PUBLISHING

ISBN 978-1-4671-2721-9

Published by Arcadia Publishing
Charleston, South Carolina

Printed in the United States of America

Library of Congress Control Number: 2017937002

For all general information, please contact Arcadia Publishing:
Telephone 843-853-2070
Fax 843-853-0044
E-mail sales@arcadiapublishing.com
For customer service and orders:
Toll-Free 1-888-313-2665

Visit us on the Internet at www.arcadiapublishing.com

To Patrick for trekking sagebrush hillsides and for technical advice,
to Eliza for dispensing knowledge,
and to Jane for moral support and tolerating it all.

Contents

ACKNOWLEDGMENTS

I am indebted to many people for their assistance. Thanks to my friend Jeff Pasek for allowing me access to the City of San Diego Water History Archive Collection (hereinafter cited as SDCWHA) that he has assembled, and for trekking to the dams in the city system. Thanks to Rick Crawford and Derrick Moses for hauling out all the Feature Histories in the San Diego City Library California Room Collection (SDCL). Thanks to Michelle Peralta and Aspen Hill at the Pioneer Room in the Escondido Library and the folks at the Carlsbad City Library; Don Smith and Jessica Sherwood at the Vista Irrigation District; Diana and Shawn Kelly at the Bonita Museum and Cultural Center; Ted Solia at Helix; Gwyneth Shoecraft of Sweetwater Authority; Jessica Parks at Santa Fe Irrigation; Robert Bergquist in Fairbanks Ranch; Mike Lee at the San Diego County Water Authority; Richard Williamson at the Carlsbad Water District; Gary Arant, Greg Hoyle, and Tony Lo Presto at the Valley Center Water District; Lesley Yeager at the San Dieguito Water District; Coleen Garcia, Laural McPhee, and Heather Smedberg in University of California San Diego (UCSD) Special Collections; Caroline Anderson of Arcadia for her patience; and the many unnamed dam enthusiasts along the way for their support.

INTRODUCTION

Regional climatic conditions determined that water in San Diego County was forever a dilemma. That dilemma became more pronounced as the municipal aspirations of the small town blossomed in the late 19th century. By necessity, San Diego water advocates assumed what Dr. Eliza Martin called the "growth by the gallon" mentality, as San Diego's city leaders pondered how best to reconcile the town's future with the water resources available. Rival Los Angeles was able to develop the Owens Valley as a water source, but San Diego had no similar alternative. To sustain a reliable water supply throughout both plentiful and drought periods, San Diego depended on the ability to collect water when it was available and store it in dammed reservoirs. The variable seasonal flow, unpredictability of the county's rivers, uneven seasonal rainfall, and cyclical droughts coupled with the region's topographic landscape determine that large-scale water storage facilities were required to systematically utilize the catchments on the county's river systems.

Private and municipal entities developed six of the seven river systems in western San Diego County. The rivers originated in the eastern interior mountains and flowed westerly through valleys that first featured steep-sided, then chaparral brush covered slopes, before they entered the tidal basins along the Pacific Coast. From north to south, the rivers were the Santa Margarita, the San Luis Rey, the San Dieguito, the San Diego, the Sweetwater, the Otay, and the Tijuana. These rivers encompassed the Hodges, Sutherland, El Capitan, San Vincente, Murray, Otay, and Cottonwood catchments, contained 24 significant dammed reservoirs, and provided the county with a combined capacity of 746,000 acre-feet of water.

The proliferation of dams on these watercourses was the region's attempt to engineer its way out of its water dilemma and attain sustainability. Private water companies constructed the first important dams. The Southern California Mountain Water Company constructed the Lower Otay Dam, the Kimball Brothers built the original Sweetwater Dam, and the San Diego Flume Company and its successor the Cuyamaca Water Company built the diverting dam on the San Diego River, Eucalyptus Dam, La Mesa Dam, and Murray Dam. In the north county, the Santa Fe Railroad and San Dieguito Mutual Water Company created Hodges and San Dieguito Dams, while the Volcan Water Company and the San Diego County Water Company built Henshaw Dam. The City of San Diego developed its independent water system through the purchase of existing private dams and then the construction of Morena and Barrett Dams in 1912 and 1923, respectively.

Dam construction in the county has continued into the 21st century, but the true golden era was between 1888 and 1934, when the county became the international nexus for dam construction. During this time, the city and private enterprise constructed the majority of the county's massive dams. But the unrestrained population growth spurred by San Diego's war economy in the 1940s overwhelmed the county's ability to supply water to its citizens and prompted the necessity of claiming water from outside resources. With the introduction of the San Diego Aqueduct, the county's oldest and largest dammed reservoirs reverted to storage facilities for the city system.

There are 50-plus dams in San Diego County that serve as storage and flood control facilities, of which 40 are publicly owned and operated. None of these dams are power generating facilities. Despite the fact that the average age of the county's dams is over 60 years, in 2016, California state inspectors judged the major dams safe. Since the arrival of Colorado River water and water from the northern Delta, San Diego County's dammed reservoirs provide 20 percent of the county's water needs and act as emergency supply facilities against the disruption of the flow from those sources.

The dams selected here are presented in chronological order to enhance the historical narrative. Many of the dams remain as integral functioning water facilities, while some only retain historical value. Civic, hydrologic, and agricultural factors determined that all these dams were located inside the western slopes of the interior mountains. Some of the monumental dams constructed between 1888 and 1934 have been heightened or had spillways reconstructed for safety concerns, but aside from the original Lower Otay Dam, which the flood of the 1916 destroyed, the dams largely stand as constructed. The county experienced a surge of construction from the 1950s through the 1980s, but given the climate of river restoration, there is little likelihood more will ever be built.

One

THE DAMS OF WESTERN SAN DIEGO COUNTY

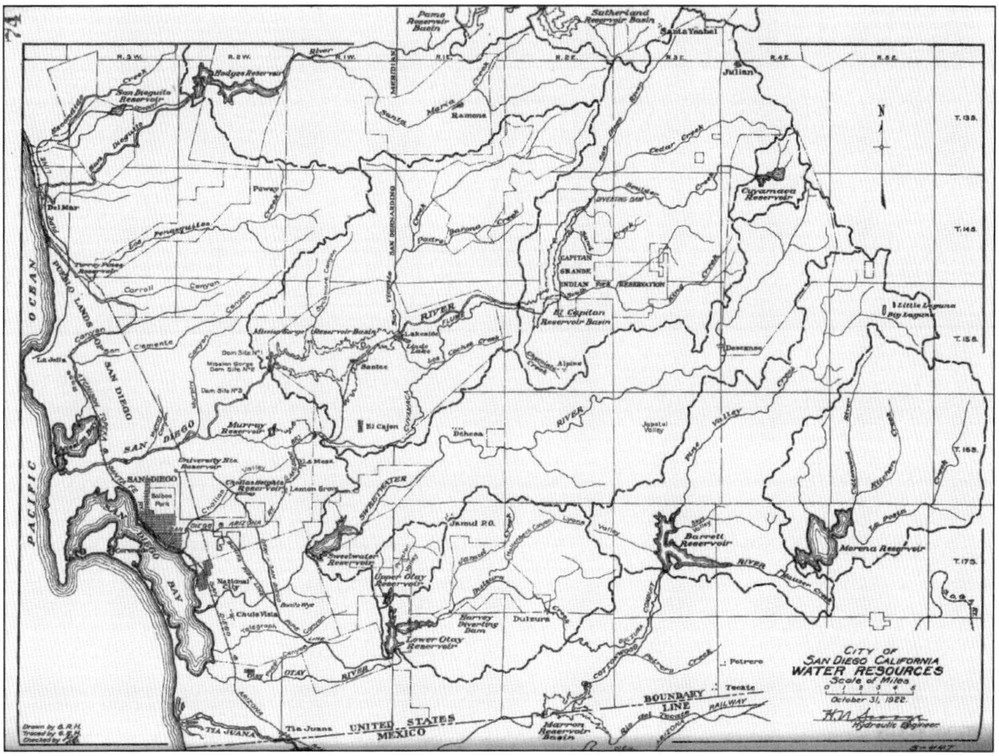

San Diego city hydraulic engineer Hiram Savage produced this map of the county's water resources at the request of the common council in 1921. Aside from El Capitan Dam, built in 1934, by this time, either private or municipal interests had constructed the county's major dams. Those dams included Sweetwater, Lower Otay, Barrett, Morena, Hodges, Henshaw, Wohlford, and Murray. (SDCL.)

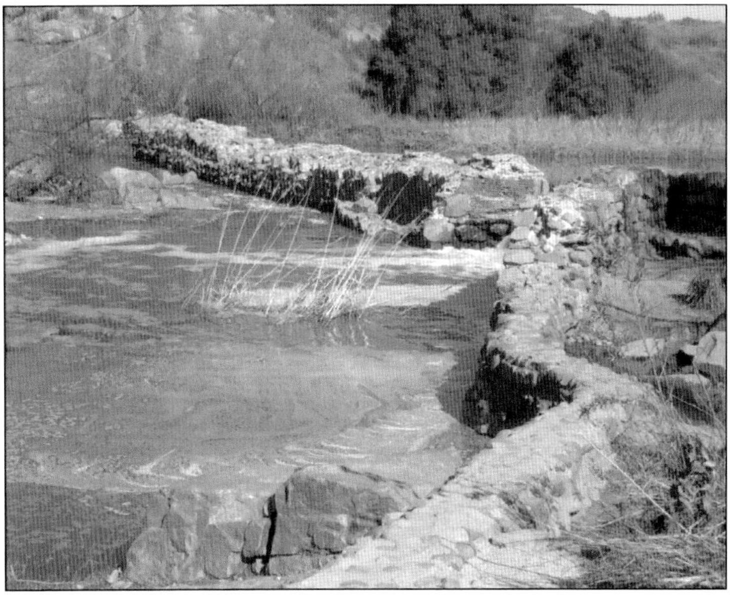

The Los Padres Dam or Old Mission Dam was both the first significant man-made water feature in the region and the first on the San Diego River. The dam also has the distinction of supplying water for the earliest major irrigation project on the Pacific coast. Locally, Mission Dam was the predecessor to all subsequent conservation dams in San Diego County. (Courtesy of Gary Van Velser.)

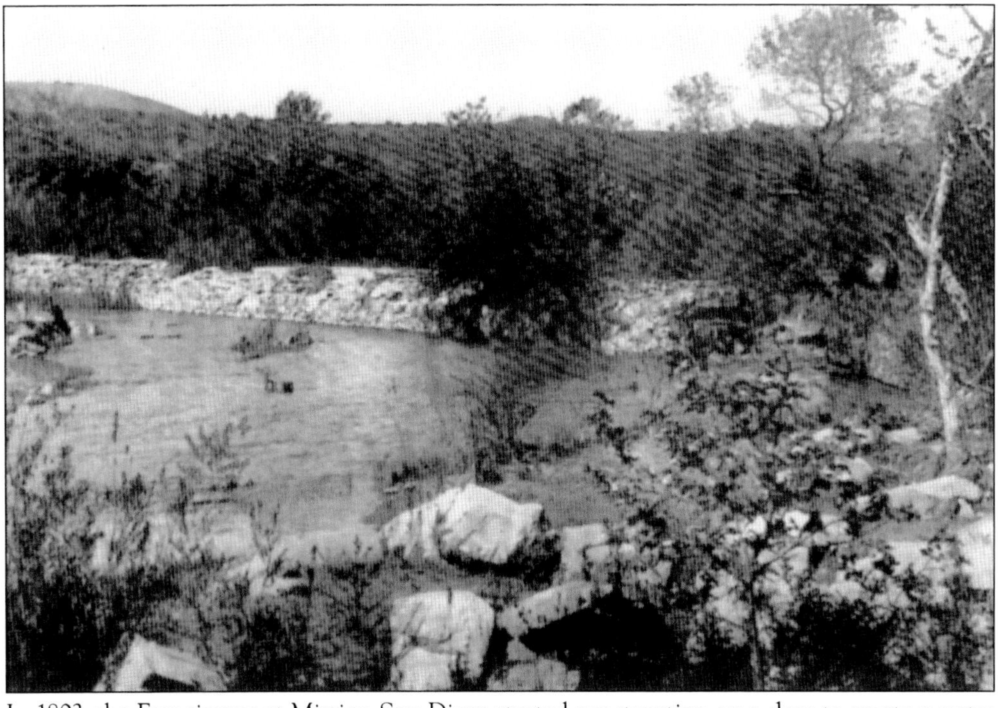

In 1803, the Franciscans at Mission San Diego started construction on a dam to create a water source to irrigate fields, water livestock, and provide domestic water. Relying on Indian labor, the fathers completed the dam in 1814. The cobblestone, brick, and concrete structure, located in a gorge about six miles to the northeast, transported water to the mission via a ceramic tile–lined aqueduct. (SDCL.)

The Mission Dam, shown here in 1917, formed a small reservoir that the mission fathers hoped would provide a water source in dry years. However, the pond, called Collins Lake or Kumeyaay Lake, was so small it dried up during a drought or when the San Diego River ceased to flow. It was over 70 years before the next major dam project in the county. (SDCL.)

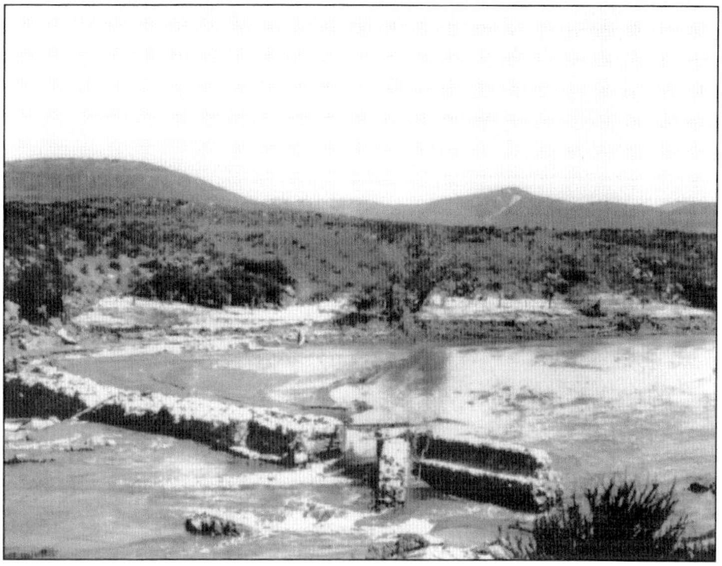

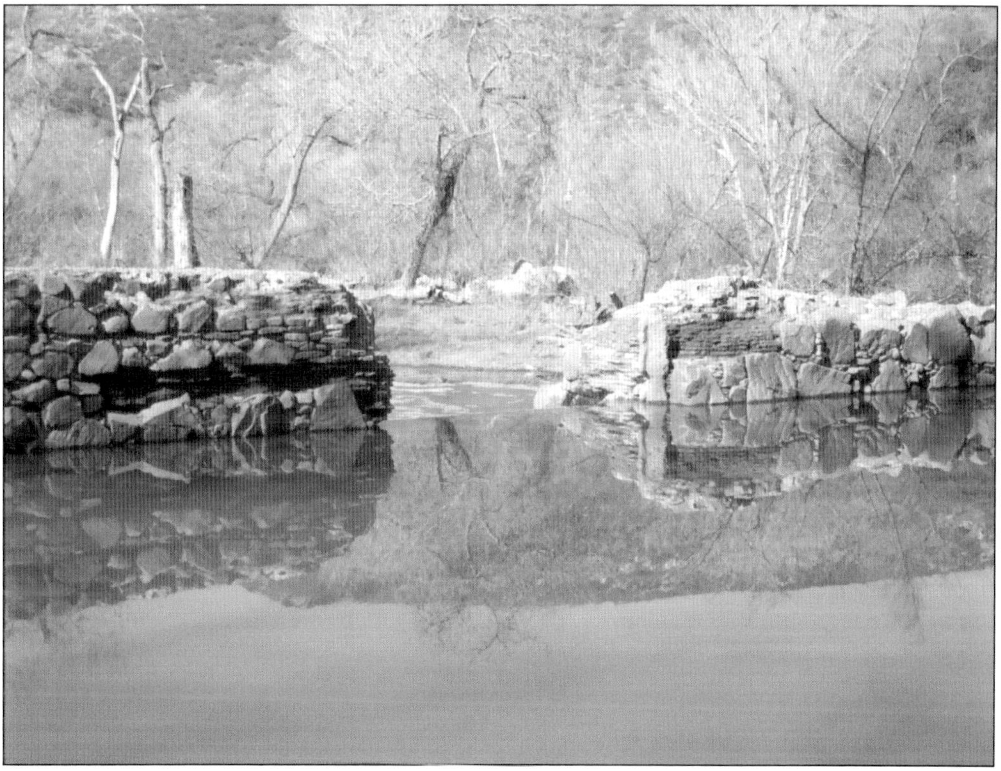

Mission Dam's usefulness diminished after the Mexican government secularized the mission and expelled the Franciscans in the 1830s. There was no motivation to repair the structure, and the dam fell into disrepair. Town leaders approved repairs to the dam in 1874, and it was briefly in use. Since 1968, the dam has been protected within the Mission Trails Regional Park as a California historical landmark. (Courtesy of Garry Van Velser.)

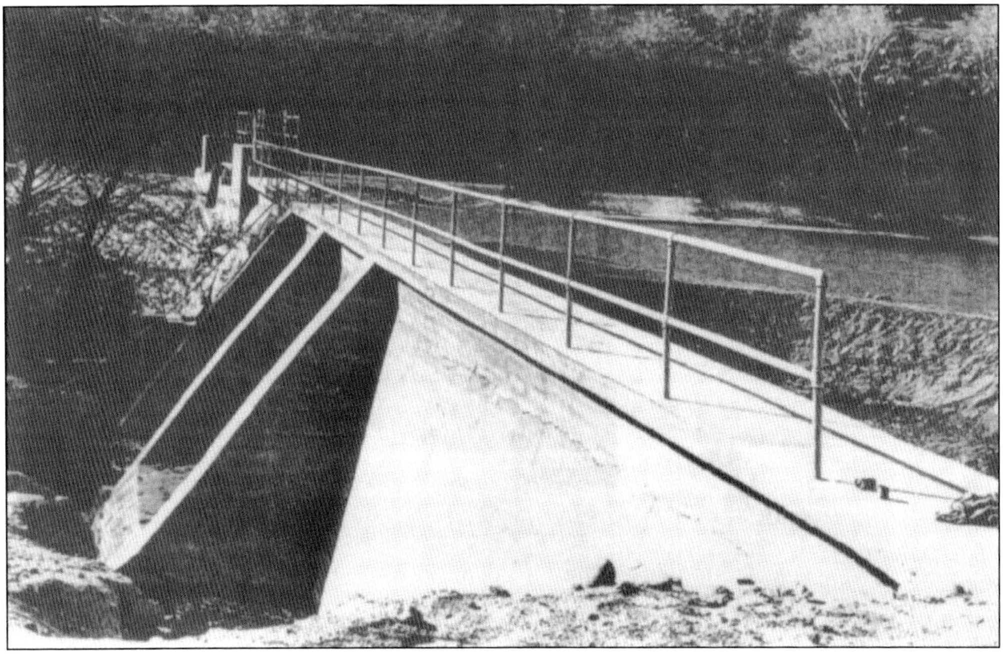

After the Mission Dam, a diverting dam on the San Diego River was the next major structure to appear in the county. In 1885, Theodore Van Dyke filed a claim for water on the San Diego River and organized the San Diego Flume Company. His plan was to take water from the Cuyamaca Mountains and the upper San Diego River and deliver it to the city through a flume system. Engineer John D. Schuyler designed a dam to divert the natural flow of the San Diego River and create a reservoir that could store around 7,000 acre-feet of water to feed the flume. Van Dyke's idea was to keep the reservoir at peak capacity and allow any excess water to flow downstream. The flume owners expended $54,400 on the original structure, which was functional, if leaky. (Both, SDCL.)

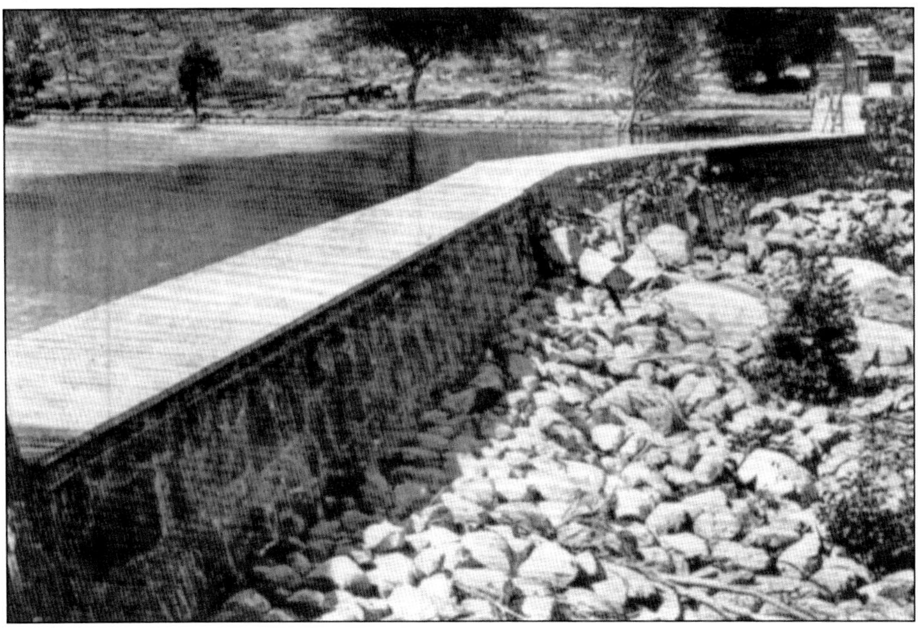

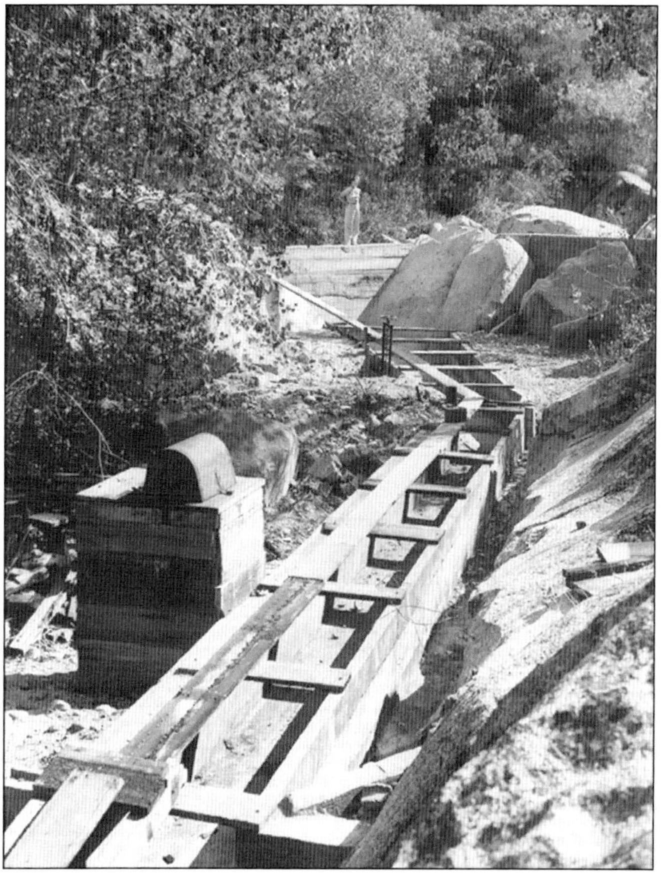

Flume officials never bestowed a formal name on the structure; it was simply known as the diverting dam. The flume engineers chose a location just below the confluence of the southward-coursing San Diego River and the outfall of Boulder Creek, where it departed a wooded valley 18 miles below Cuyamaca Lake. Water was released directly into the flume system through a manually operated gate at the base of the east end of the dam. At 810 feet in elevation, the water gravity-flowed out of the 35-foot dam through the flume to the town of Lakeside and on to a catch site before being relayed to the city distribution system. (Both, courtesy of Helix Water District.)

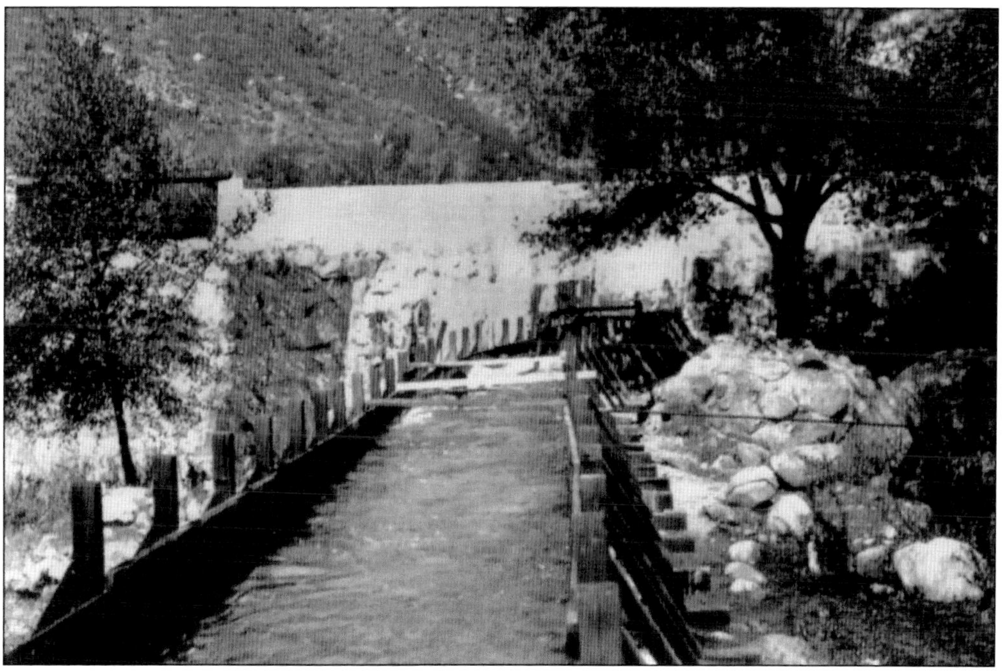

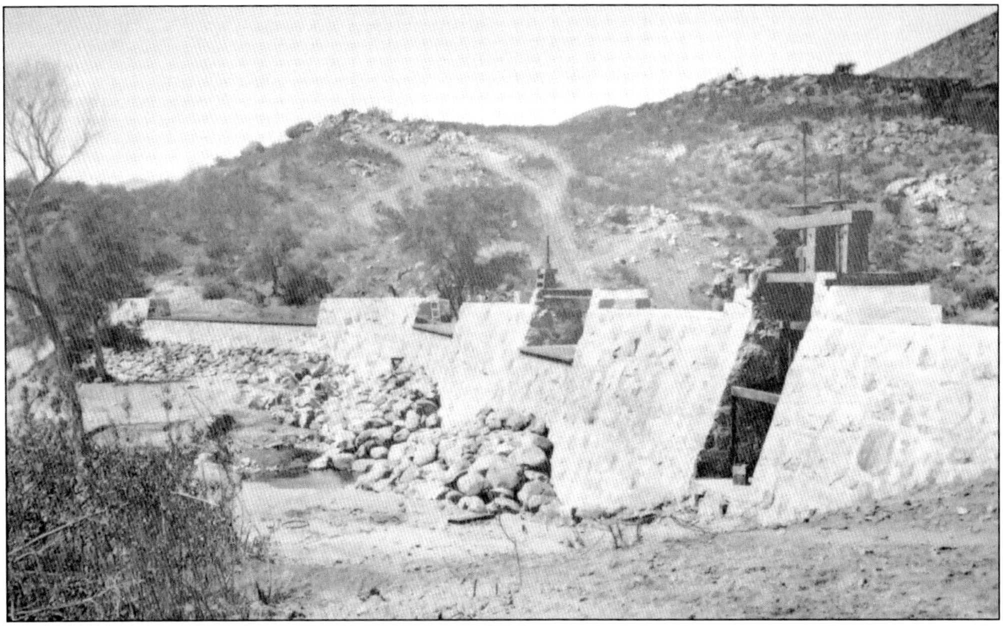

The original dam, shown in this 1897 photograph, was a 540-foot-long stone and masonry earth-filled dam with a simple overflow channel spillway on the west end and overflow points near the middle of the flat crest. After completing the dam and testing its integrity, the engineers decided to strip off the original rip-rap face and apply a smoother concrete and masonry layer. (SDCL.)

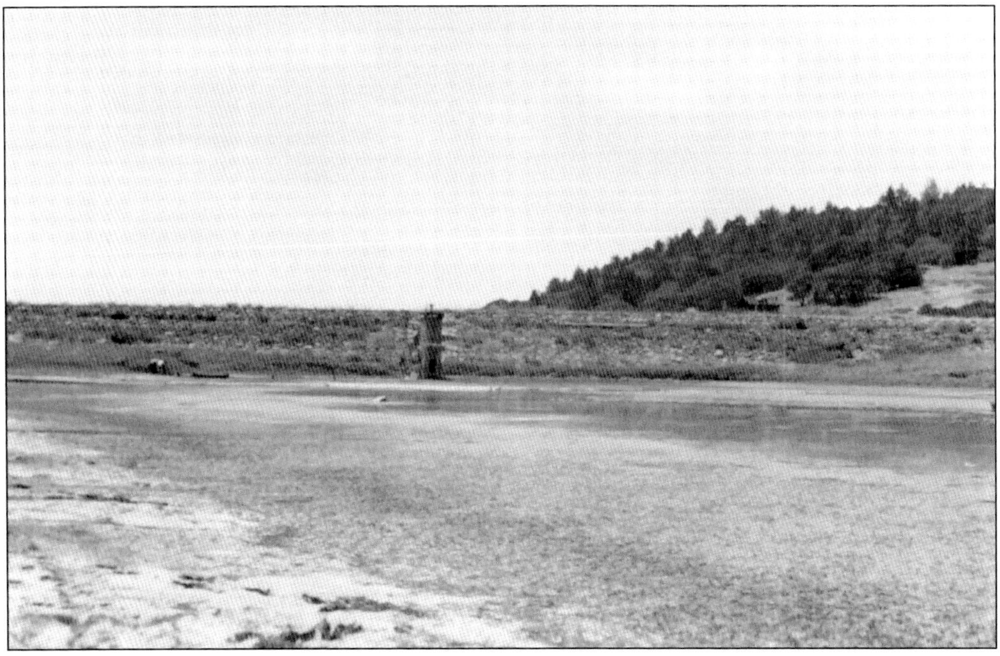

For Van Dyke's flume system to be successful, it required more water than the seasonally flowing San Diego River could supply. To supplement the supply, he planned to build a reservoir in the surrounding Cuyamaca Mountains, which received an annual rainfall of 36 inches. Van Dyke sold the idea to a group of British stockholders, and the San Diego Flume Company became reality. (SDCWHA.)

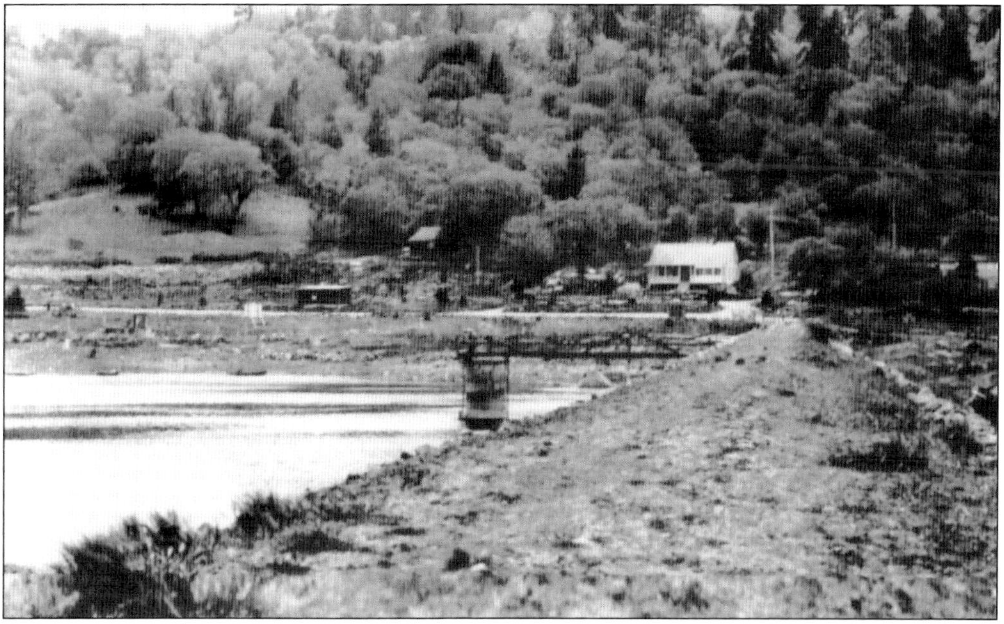

In 1888, Van Dyke purchased a site at a natural pond at the headwaters of Boulder Creek, an eastern tributary on the upper San Diego River, for $54,000. The British consortium funded the construction of Cuyamaca Dam there that September. The timing of the enterprise was unfortunate, as a 10-year drought descended on San Diego from 1895 to 1905 and the river and the reservoir dried up. (SDCWHA.)

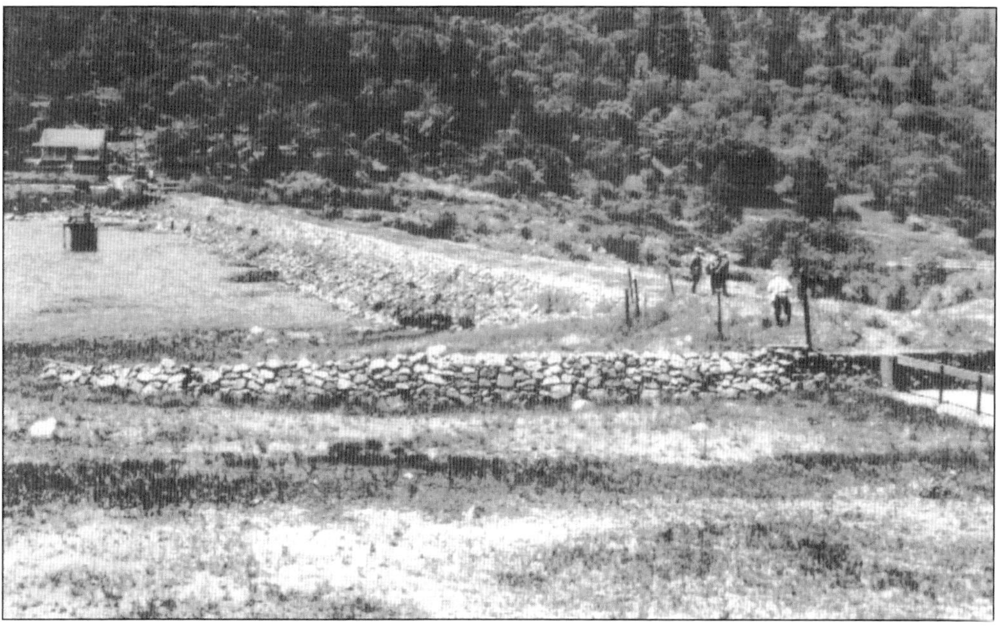

Cuyamaca was originally to be a masonry dam, but the lack of a quarry necessitated constructing a low, earth-filled structure. Completed in February 1887, the low sloping dam stood 35 feet in height and 635 feet in length, with a 10-foot-wide crest. The flume owners called the area Cuyamaca, but the Indians called it Laguna Que Se Seca—the lake that dries up—for reasons the flume owners soon discovered. (SDWHA.)

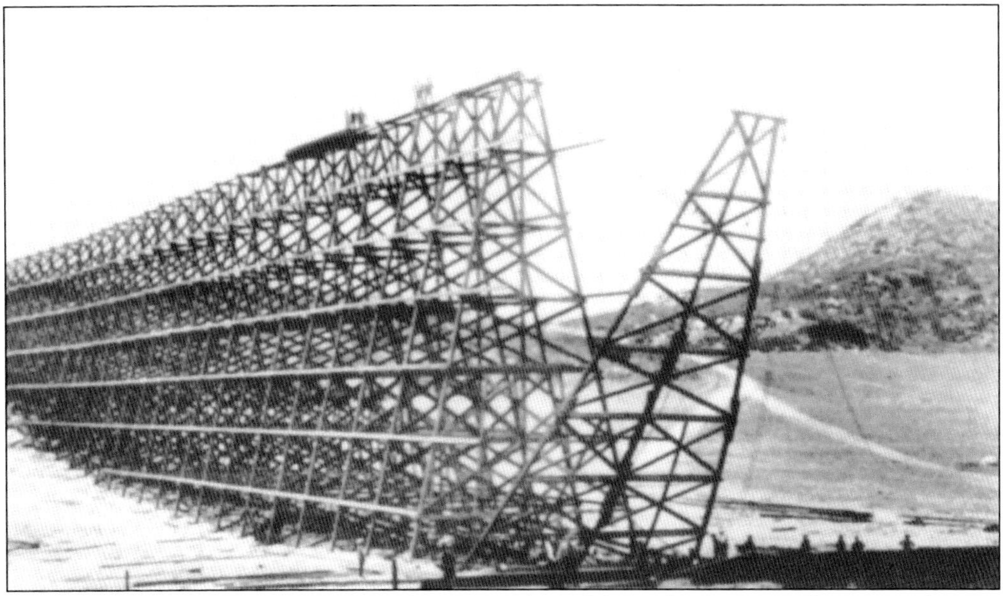

The flume was an engineering feat. The water released from the diverting dam poured into a redwood chute that transported it over 30 miles downstream. Roughly paralleling the San Diego River, the open wooden trough carried the water over raised sections, through tunnels, and across high wooden trestles, with distribution points for the backcountry ranches and farms along the river valley. The flume was a remarkably leaky structure, and drought and the problematic nature of the upstream water source rendered the system unreliable. However, despite debris and dead animals clogging the flow, and the recreation value for ranchers' kids, the flume was in service until the 1940s. The flume's final destination was Eucalyptus Reservoir on Grossmont Knoll. (Both, SDCHWA.)

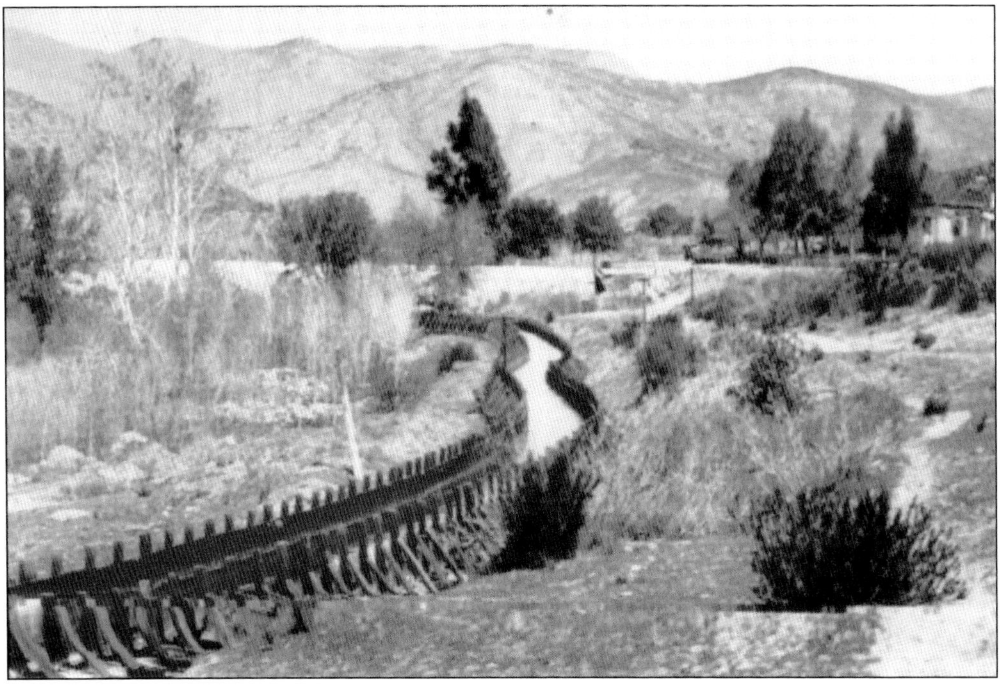

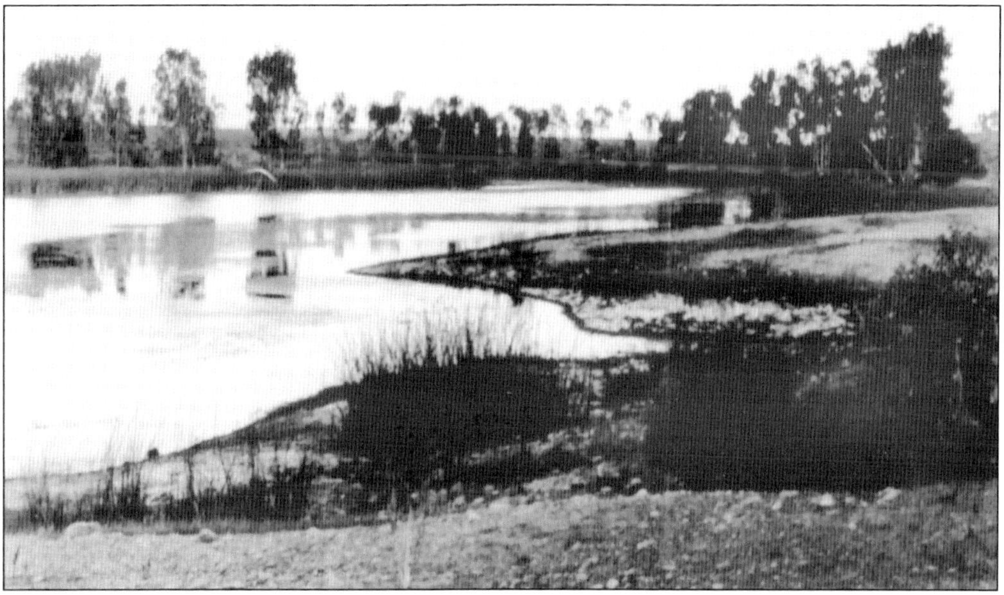

Eucalyptus Dam and Reservoir, located near Grossmont Heights, was the flume's termination point. The 34-foot-high, 275-foot-long dam that conserved the water was apparently named for the abundance of tall non-native eucalyptus trees that fringed one end of the shallow reservoir. The small dam had a well-traveled dirt road across its crest and a metal walkway connecting the dam crest to the outlet tower. (SDCWHA.)

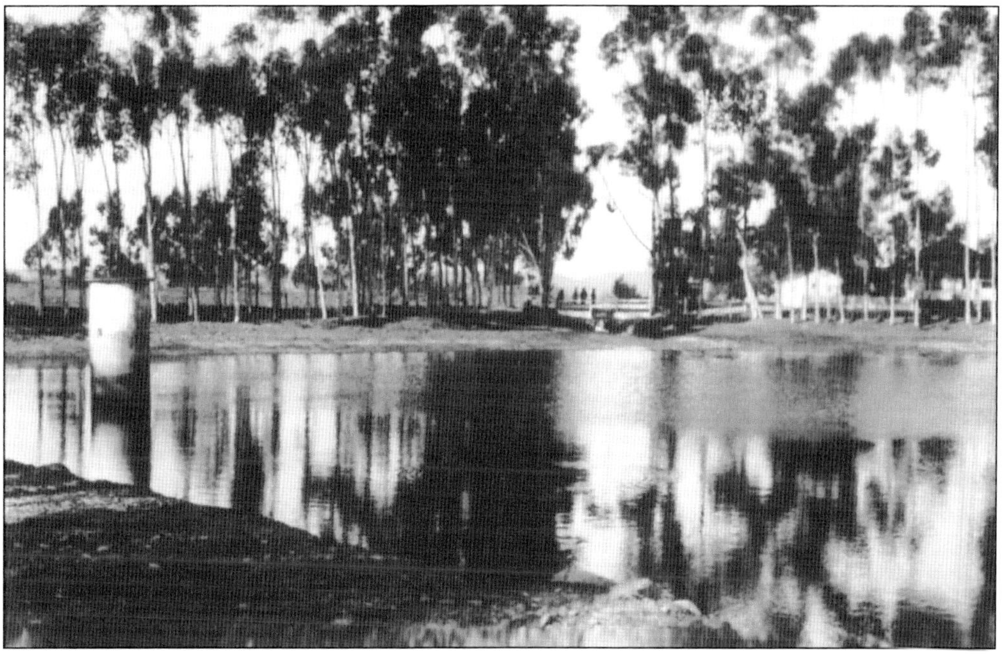

Eucalyptus Reservoir was a storage or regulation facility. The flume managers moved the water as needed to La Mesa or Grossmont Reservoir. Supplying water to San Diego was the most lucrative water concession in the county, and the flume managers depended on a constant flow of water into the Eucalyptus facility to create a ready supply for city consumers. If the flume went dry, so did Eucalyptus Reservoir. (SDCWHA.)

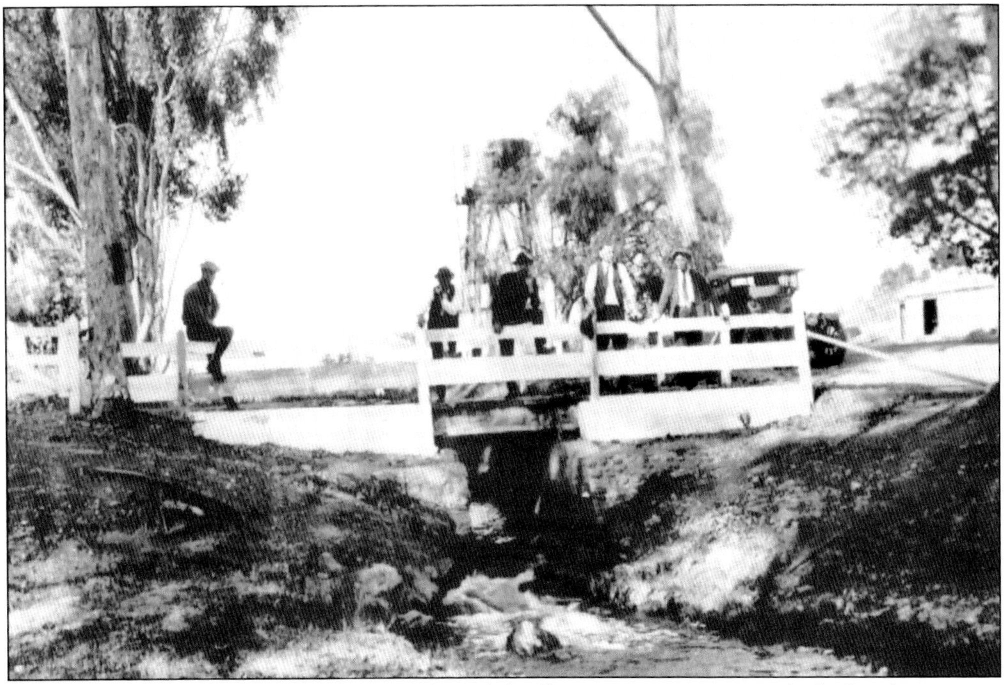

The Eucalyptus Reservoir outlet tower released the flume water into an earth ditch, where it was siphoned across a deep ravine in a 3,000-foot-long, 36-inch redwood stave pipe. The pipe emptied into the mile-and-a-half-long La Mesa Ditch, which discharged the water into the south side of La Mesa Dam. La Mesa Reservoir stored the water until it entered the city system. (SDCWHA.)

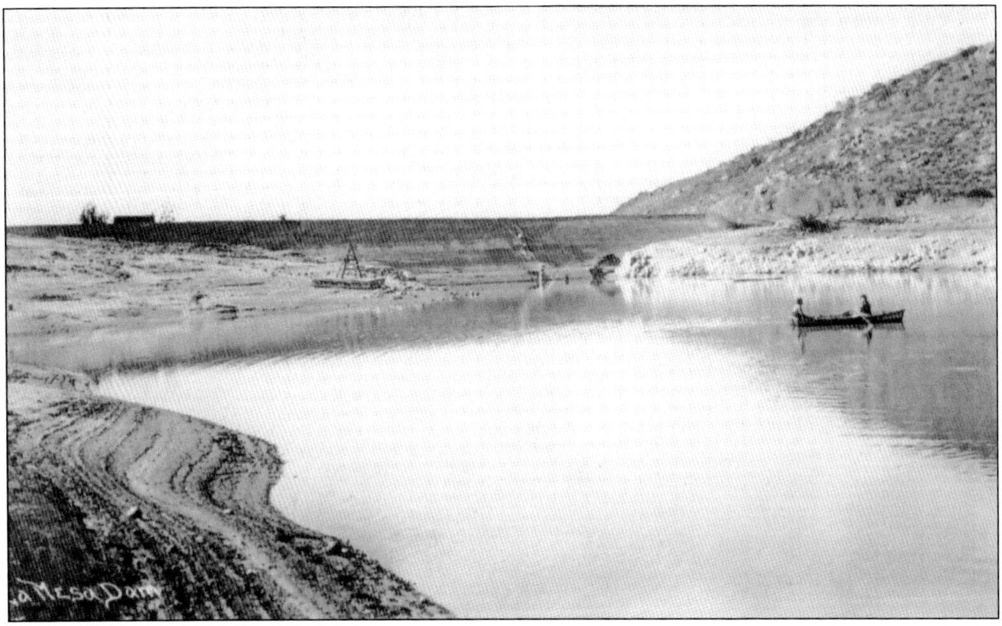

After the two-mile trip from Eucalyptus Reservoir through a series of pipes, conduits, and the La Mesa Ditch, the water reached its termination point, the loose rock–fill hydraulic La Mesa Dam, located just northeast of the town of La Mesa. The dam spanned Alvarado Canyon and conserved the Alvarado Creek, a tributary of the San Diego River. (Courtesy of Helix Water District.)

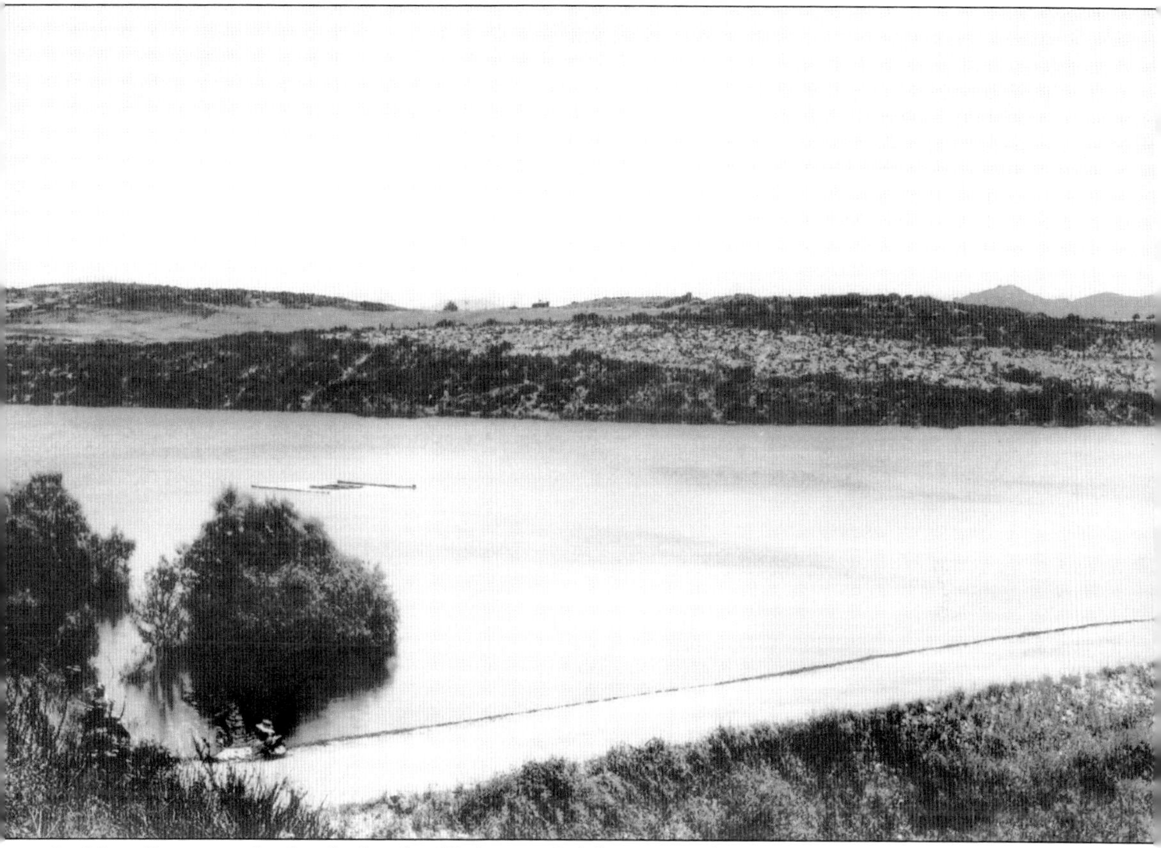

La Mesa Dam was the first hydraulic fill dam in California. Workers completed the 66-foot-high, approximately 500-foot-long dam in 1895 at a cost of $17,000. From its completion, La Mesa was a small, porous structure. Whether it was the concept or inferior construction, over 23 gallons per minute seeped from the earthen dam when the water height reached 46 feet, and the dam lost over 100 gallons per minute at 54 feet. In an attempt to arrest the leakage, engineers covered the upstream water slope with a facing of what was then called asphaltum cement concrete. The spillway was an open channel along the left abutment. The Cuyamaca Water Company purchased the flume system in 1910 and made plans to replace La Mesa with a larger structure. The dam escaped the flood of 1916, but even with the small catchment, the floodwaters almost overtopped it. The new Murray Reservoir slowly submerged La Mesa Dam, which is now only visible during extreme drought conditions. (Courtesy of Helix Water District.)

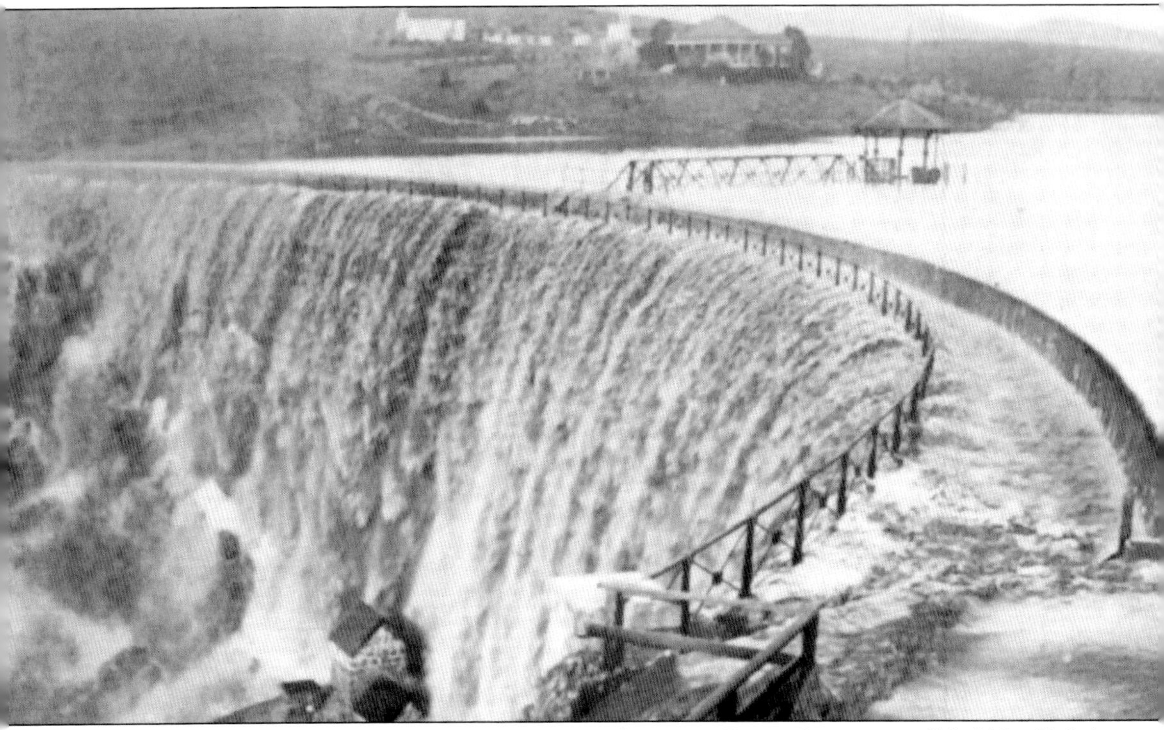

The construction of the Sweetwater Dam in southern San Diego County paralleled Van Dyke's efforts to establish the San Diego Flume Company. Along with Cuyamaca Dam, the Sweetwater Dam is the oldest continuously functioning structure in San Diego County. The Kimball Brothers Water Company initiated the first effort to conserve the Sweetwater River, which coursed through the old Mexican land grant called Rancho de la Nacion. The Kimball brothers, Frank, Levi, and Warren, selected the Sweetwater River because they believed it was one of the few rivers in the county that flowed year-round and could sustain their agricultural lands and also supply domestic water to National City and Chula Vista. Using plans from engineer Frank E. Brown, workers laid the foundation for a 90-foot-high dam in 1888. This US Geological Survey photograph shows the dam overflowing during a flood event in the 1890s. (SDCL.)

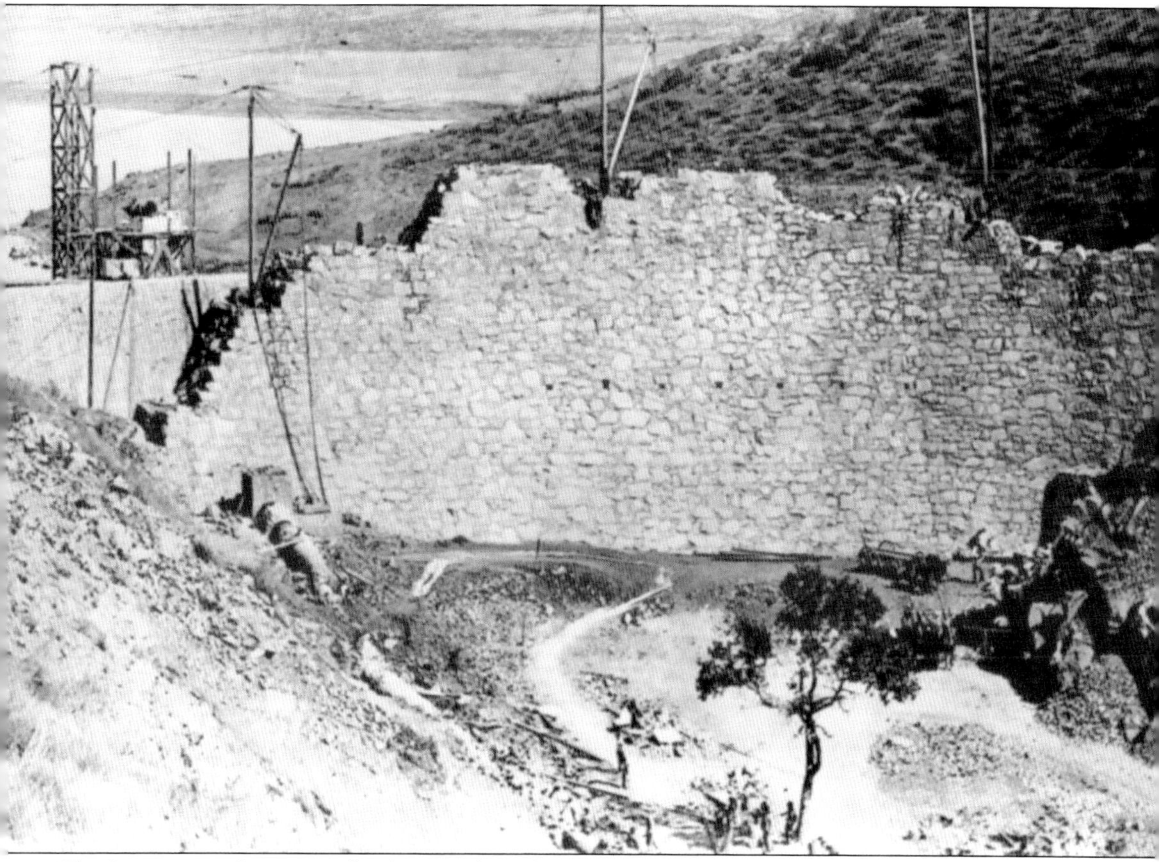

The Sweetwater Dam was the first modern dam in San Diego County, and at the time of its dedication, it was the highest masonry arch dam in the United States. The dam suffered through several design changes before the San Diego Land and Town Company took over the project in 1886. The dam was a concrete gravity arch structure located about 12 miles south of San Diego and directly inland from the town of National City. The dam's downstream face was lined with 36-foot-by-15-foot stone blocks hauled from a nearby quarry and set with concrete. The reservoir immediately suffered some serious leakage issues. In an effort to mitigate the problem, but ignoring the assistance of professional engineers, the builders lined the bottom with a 10-to-15-foot-thick "blanket." It soon became apparent that the dam was insufficient in size and perhaps even structurally unstable. (SDCWHA.)

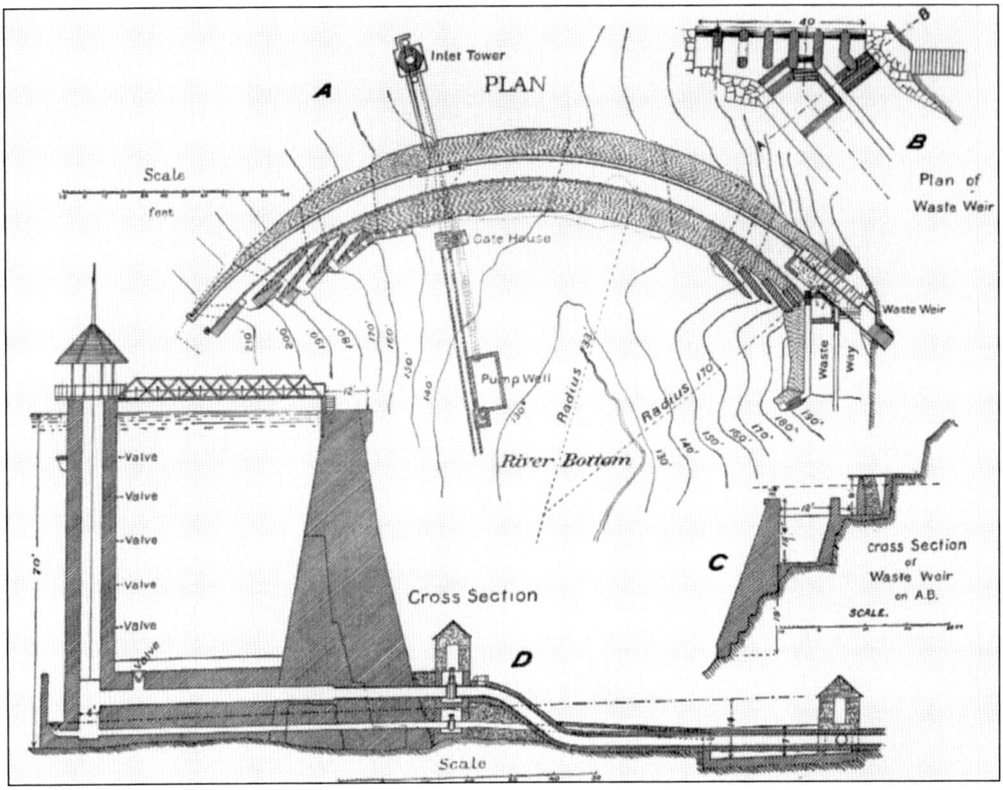

When the San Diego Land and Town Company managers took control of the Sweetwater project in 1886, they hoped the reservoir would supply the domestic use and irrigation water for their planned real estate development near National City. The dam, completed in 1888 after only 16 months, cost $234,074. Sweetwater stood 108 feet high, was 700 feet in length, and created a reservoir with 960 acres of surface area and a 28,779-acre-foot capacity. Construction time was shortened when the National City & Otay Railway completed a spur line to the dam site, which eased the movement of materials. The spillway consisted of eight overflow channels on the south side of the dam with concrete abutments to control the flow. Engineers retrofitted the dam several times between 1888 and 1890 but never properly addressed the spillway's capacity. (Both, SDCWHA.)

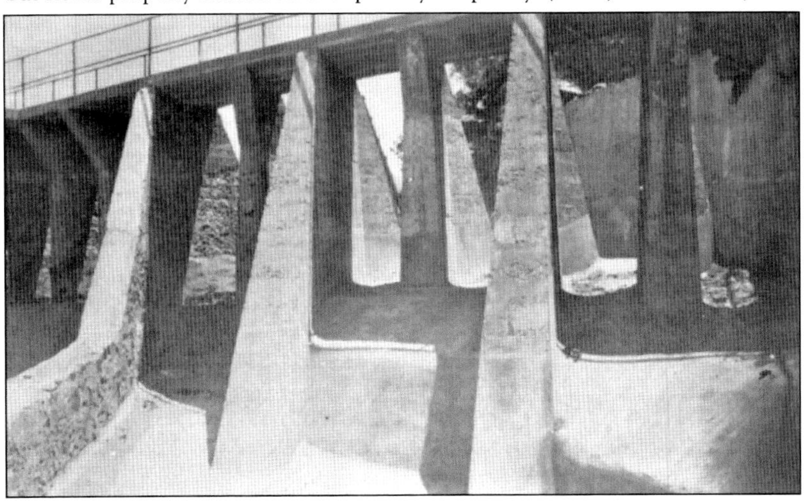

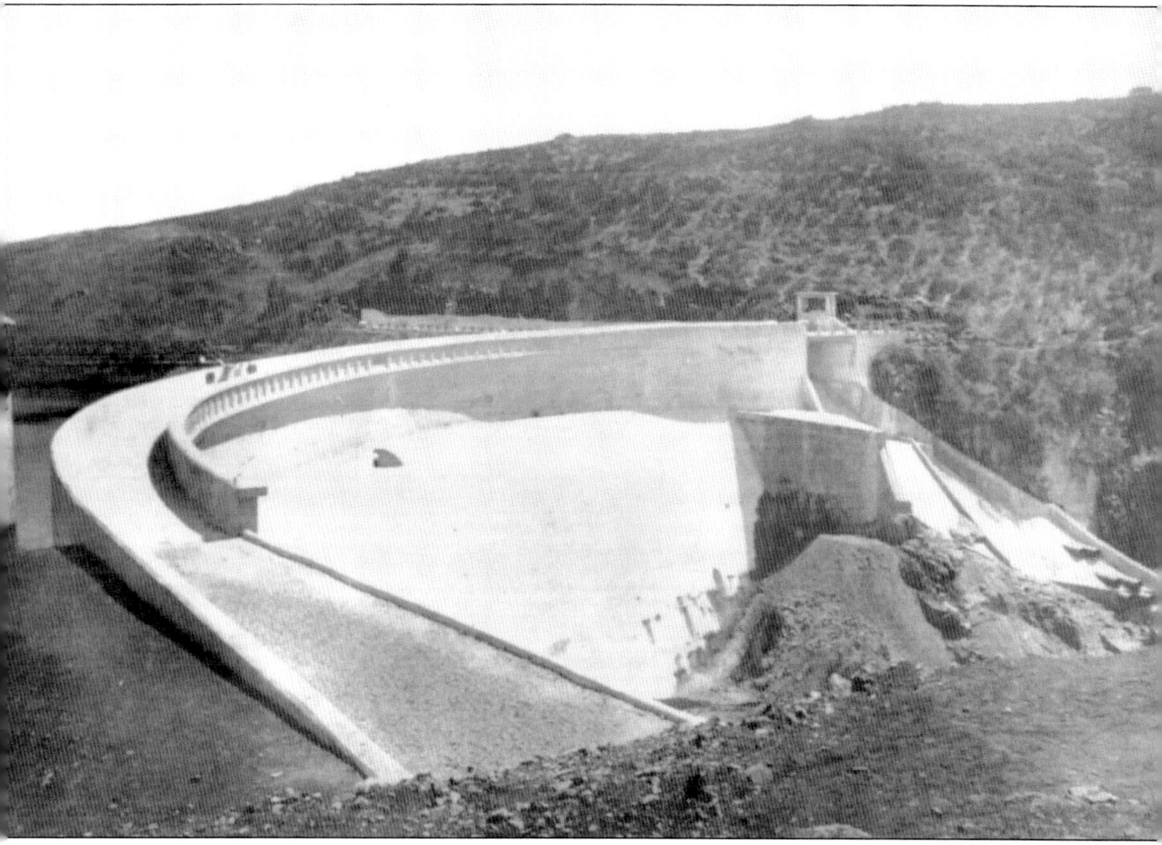

Flooding on the Sweetwater River in 1895 slightly damaged the dam, and the subsequent drought period between 1898 and 1904, which left the reservoir virtually empty, negated Kimball Brothers' belief in the river's reliability. Their trust was tested between 1898 and 1904 when the reservoir went dry, and again in 1906 when it was only half full. The drought ended in 1909, and seasonal flooding filled the reservoir to overflow. However, those successive years of drought motivated the sale of the dam to John Spreckels's Southern California Mountain Water Company (SCMWC) in 1910. The new owner increased the height of the dam to 110 feet, reconstructed spillways, and raised the intake tower. State engineer John D. Schuyler supervised the work to raise the dam's crest height to 112 feet and place a concrete "saddle dam" on the south rim, as shown in this 1911 photograph. (SDCL.)

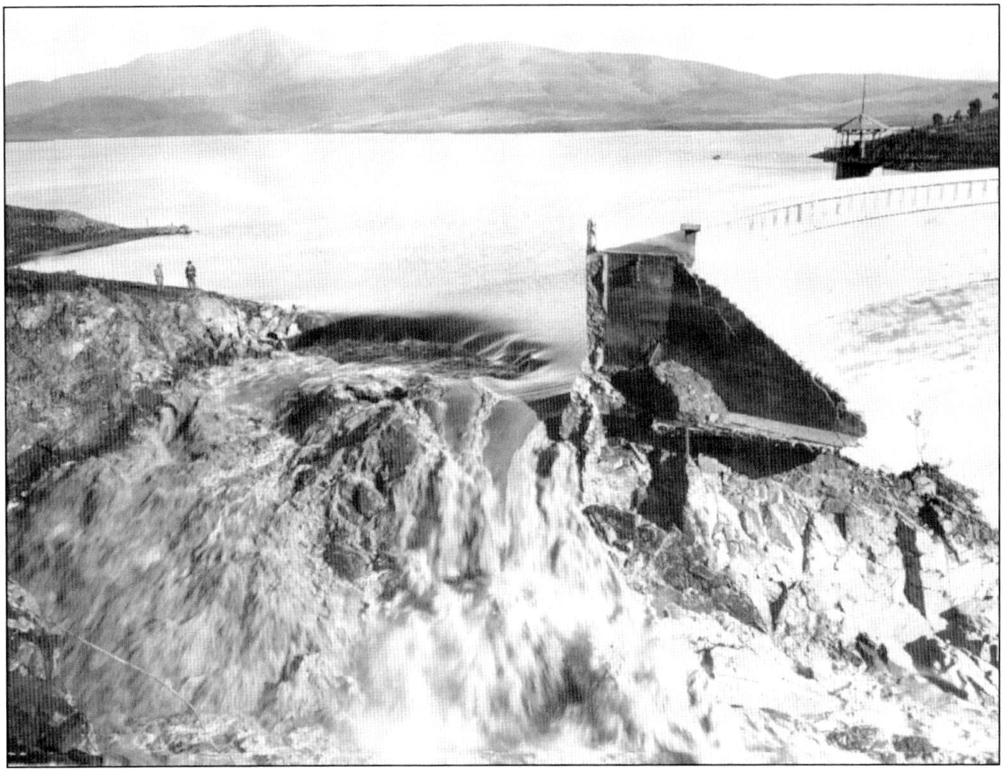

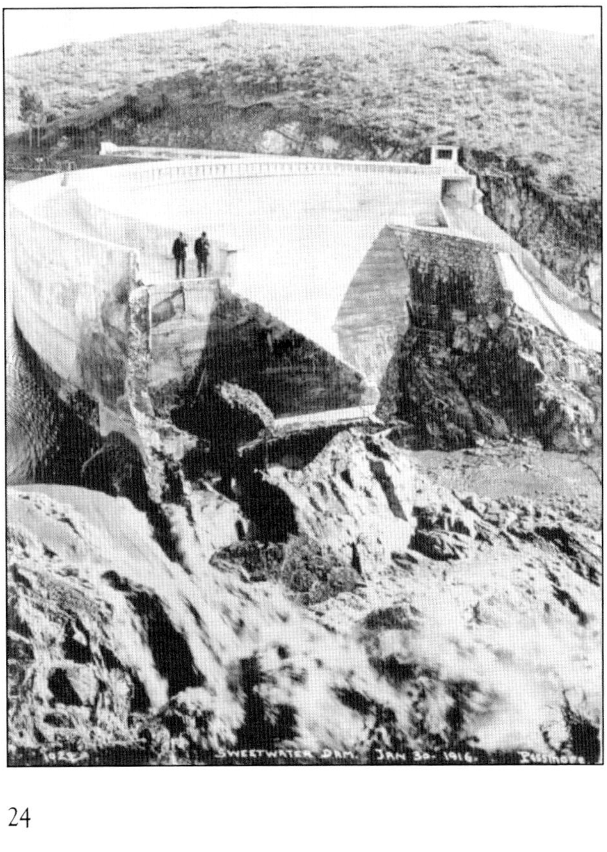

The flood of 1916 revealed the flaws in Sweetwater's construction, specifically the inadequacy of the spillway. In January, surging floodwaters overtopped the structure, destroyed the earth embankment on the north abutment, and forged a 90-foot-wide gap that emptied two thirds of the reservoir. The failure sent a torrent of floodwater through National City and into San Diego Bay. The floodwaters destroyed downstream pipelines, railway tracks, private property, and utility poles as they swept across the river valley. Fortunately, the flood spared the main structure of the dam. With the majority of the dam's face intact, a reconstruction project was viable. (Both, SDCL.)

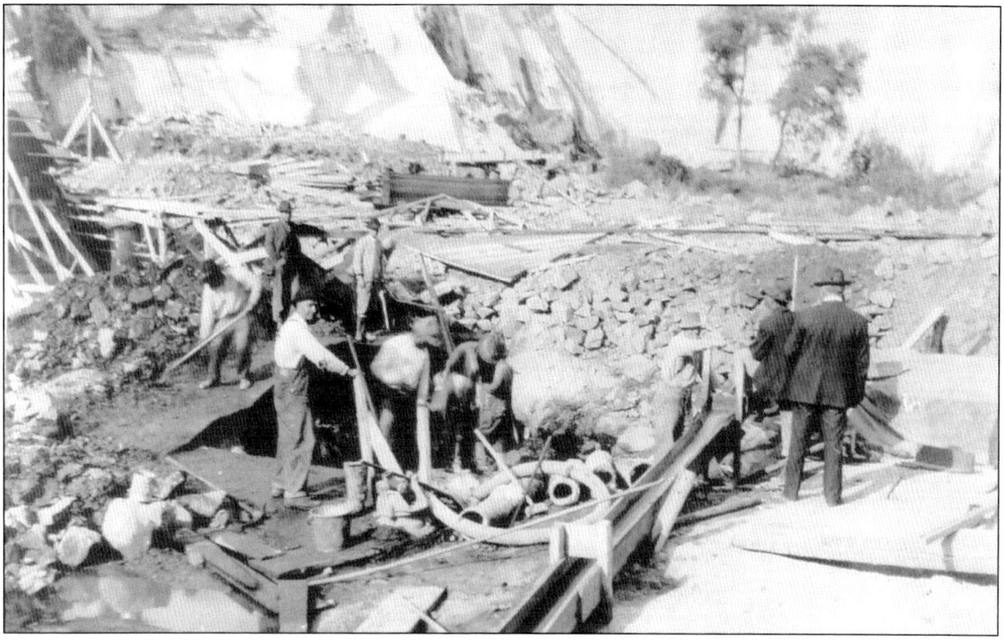

San Diego city hydraulic engineer Hiram Savage, shown above inspecting the progress at Sweetwater in 1917, took advantage of the reconstruction to address the inadequate spillway. Savage designed a massive siphon spillway on the left side of the dam. The new spillway housed six 12-foot-wide barrels with new parapets above them. Cleverly, Savage placed a stilling weir at the base of the dam, which dissipated the energy of water passing over the crest and protected the base of the dam from erosion. The dam was overtopped in the 1927 flood but suffered no serious damage. In 1939, the dam underwent a major overhaul. The dam is shown below in 1917 after the completion of the new spillway. (Both, SDCL.)

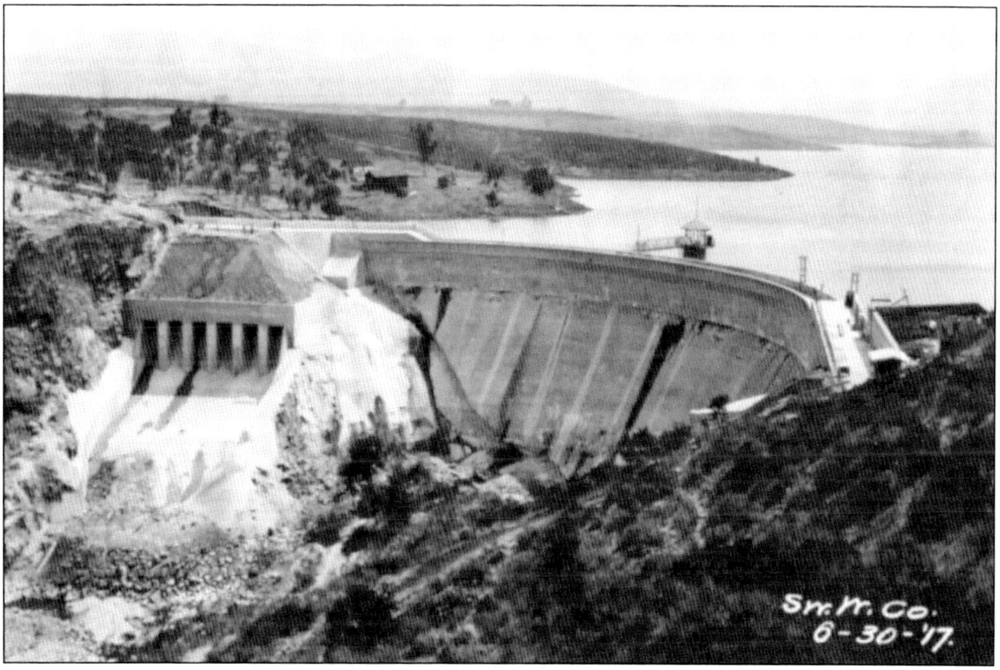

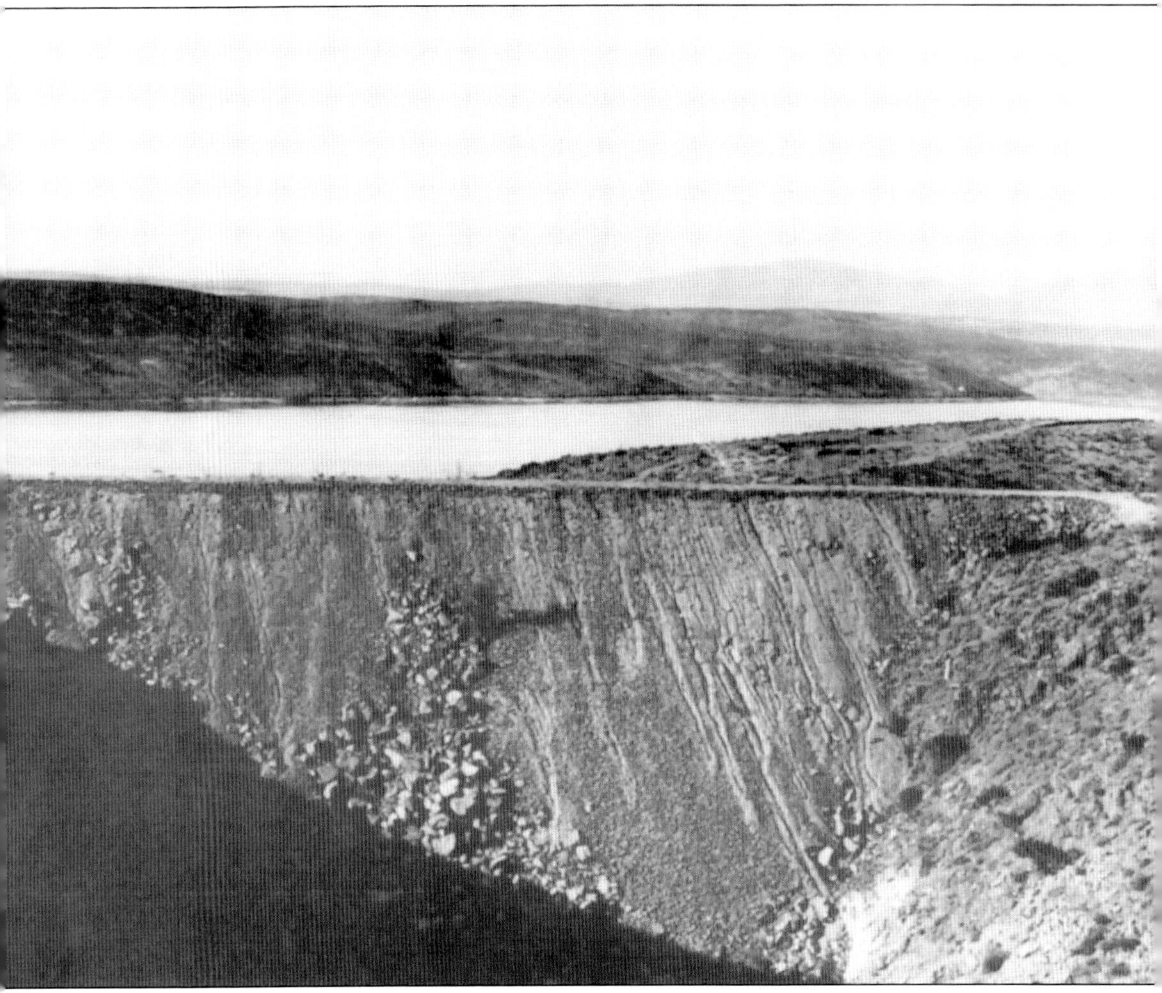

The Otay Water Company started work on the masonry Lower Otay Dam in 1886, two years before the completion of the Sweetwater Dam. When work on the dam faltered from funding and construction issues, Otay Water sold the site on the Otay River to the SCMWC. Amateur engineer Elisha S. Babcock, one of the original investors in the Hotel del Coronado, assumed the work for the new owners and supervised the work to completion in 1897. A self-trained engineer, Babcock's construction methods were largely unconventional and perhaps naive in their application. His plans for the earth-fill concrete structure included the unique idea of a central wall of vertical steel plates within the exterior fill to ensure water tightness. The steel plates rose to within three feet of the top of the fill section, with some of the sheets 50 feet tall. Lower Otay Dam, located about 20 miles southeast of San Diego, stood 125 feet high by 565 feet long. (SDCWHA.)

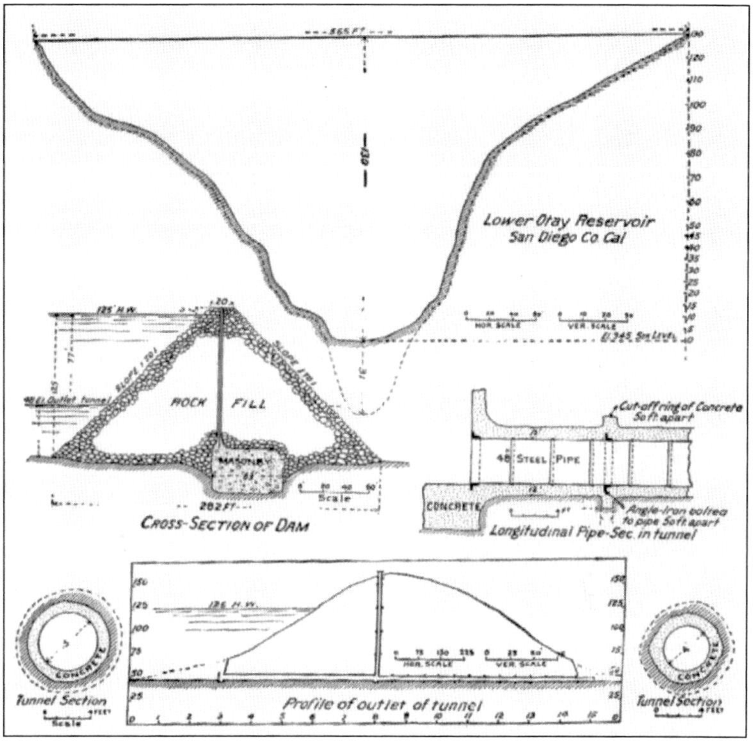

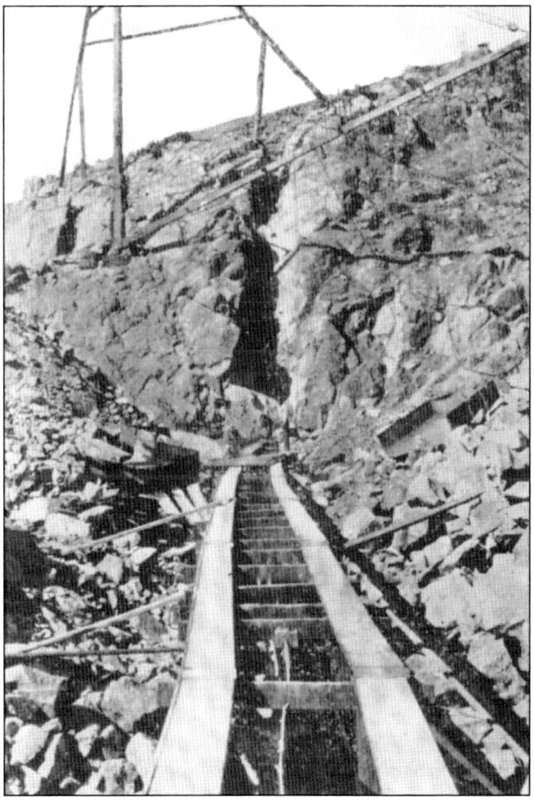

Following Babcock's plans for Lower Otay, workers riveted the steel plates in place, caulked the seams, and coated both sides with a layer of burlap saturated with hot asphalt. Supposedly the burlap would prevent the coating from running and create a smooth, even, uniform surface. Laborers applied a second coat of asphalt to form a hard facade that would hermetically seal the entire surface of the plates and prevent the steel from rusting. An interior wall was constructed to stiffen the structure and protect the steel plates against possible damage when rock was dumped around them. Babcock designed and built a masonry wall, six feet at the base tapering to one foot, on each side of the plate. To further bolster the dam, workers slotted the plates inside trenches they cut into the solid rock of the canyon walls. (Both, SDCL.)

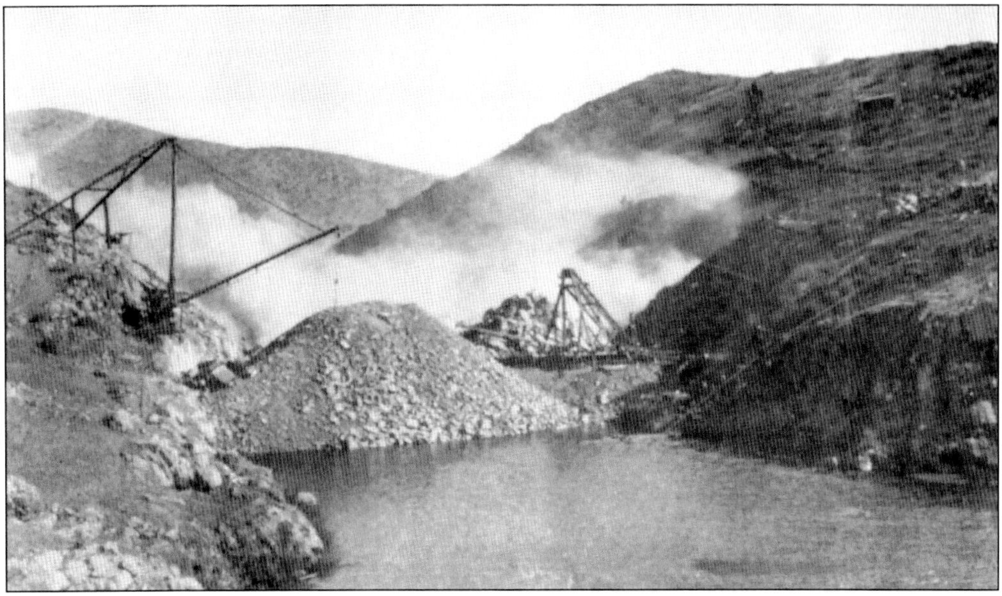

The workers at Otay used a Liderwood cableway to move rock blasted into manageable pieces at the quarry to the construction zone. The cableway had a two-and-one-quarter-inch-diameter cable with a 980-foot span and a capacity of 10 tons. The cableway moved rock to the derrick, which dropped the stone where men could distribute it with a secondary cableway system or handcarts. (SDCL.)

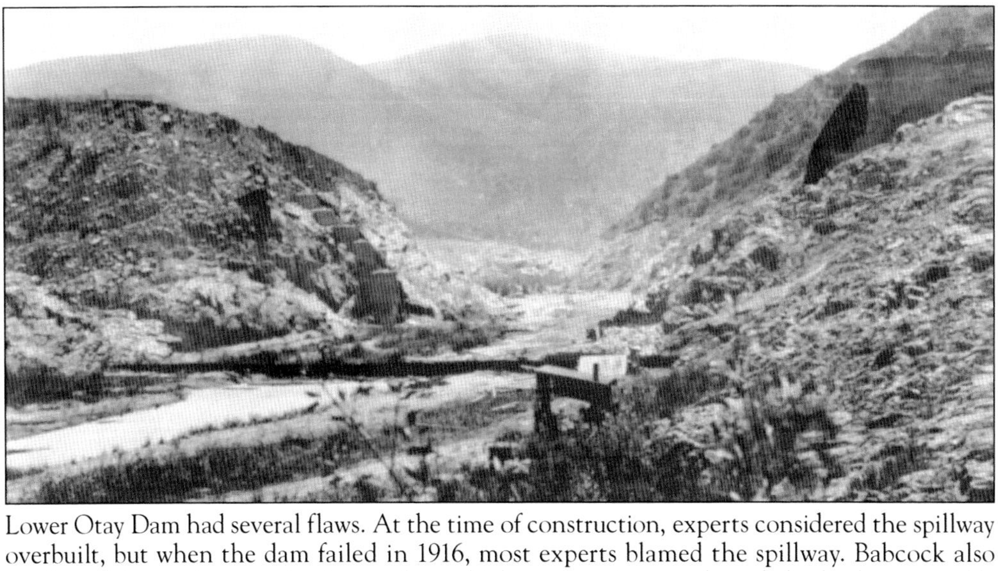

Lower Otay Dam had several flaws. At the time of construction, experts considered the spillway overbuilt, but when the dam failed in 1916, most experts blamed the spillway. Babcock also configured Otay to be an embankment dam, which many engineers acknowledged to be susceptible to erosion, and made the upstream and downstream slopes steeper than ordinarily used for an embankment dam. The result was ultimately disastrous. (SDCL.)

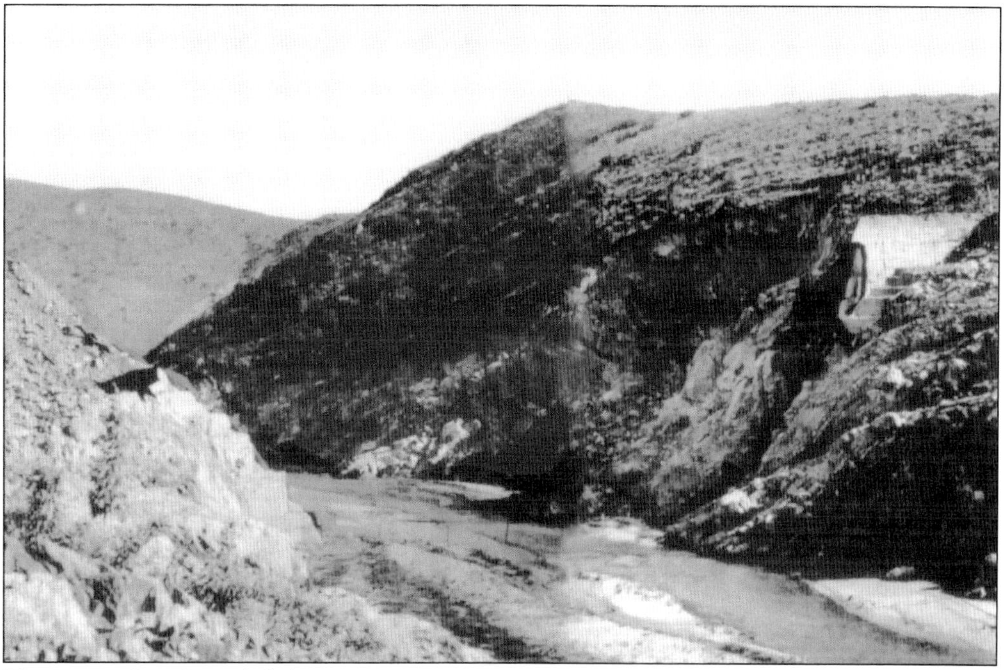

The Lower Otay Dam failed in January 1916 when the rains from consecutive storms expanded the reservoir capacity by 9.5 feet in 10 hours. Fifteen minutes after the dam overtopped, the structure virtually disappeared, releasing 40,000 acre-feet of raging water into the Otay Valley. The runaway waters destroyed residences and ranches, killing 14 people. The failure motivated California lawmakers to enact new legislation for dam construction and safety. (SDCL.)

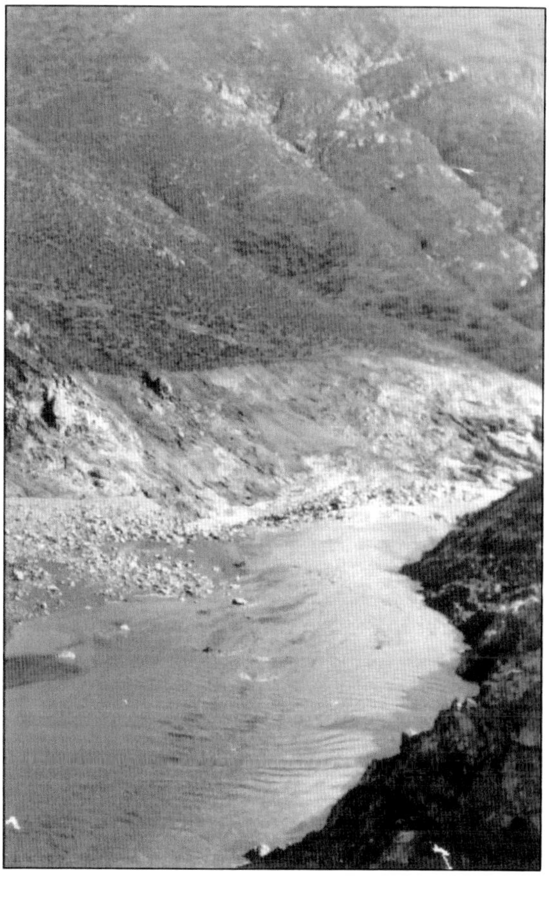

City officials blamed the Otay disaster on so-called "Rainmaker" Charles Hatfield. The episode started when the city agreed to pay Hatfield $10,000 in late 1915 to conjure a rain event and fill the new Morena Reservoir. Largely for liability issues, after the disaster, the city blamed Hatfield and refused to pay. Hatfield sued the city but despite years of litigation never received any payment. (SDCL.)

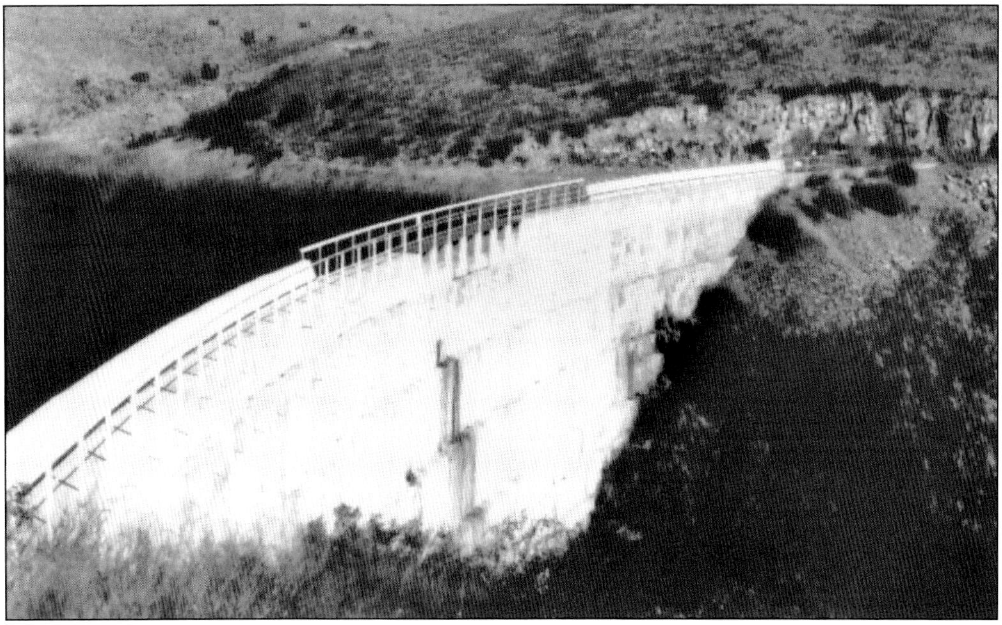

As Babcock worked on the Lower Otay Dam, he also designed and supervised the construction of the nearby Upper Otay Dam for the SCMWC between 1896 and 1898. Locals applied the name because of the dam's location immediately above Lower Otay Dam. The dam conserved Proctor Creek, a tributary of the Otay River, and was built to create a reservoir with 2,800 acre-feet of capacity. (SDCWHA.)

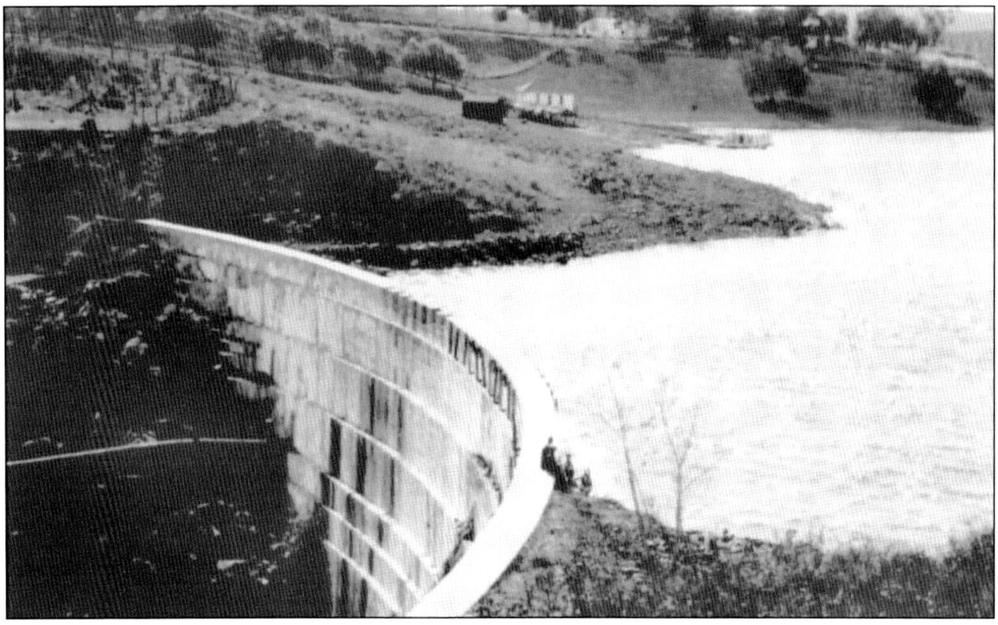

The original dam was a 24-foot-high masonry structure, but Babcock quickly realized that a larger dam was necessary to make the project viable. Ultimately, the dam reached a height of 84 feet. The structure was a thin concrete constant-radius arch dam that extended 353 feet and had a narrow four-foot crest. The final project cost was $80,000. Although overtopped, the dam survived the flood of 1916. (SDCL.)

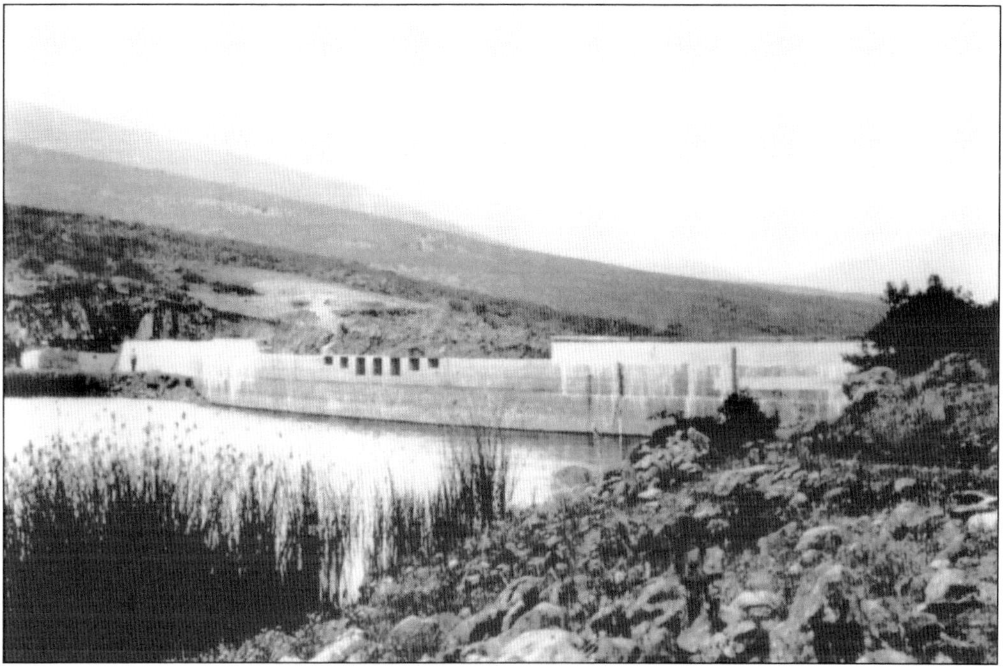

As at Lower Otay, Babcock used steel plates at Upper Otay to reinforce the very thin arch, which was only 14 feet at the base. Builders riveted two heavy steel plates together, placed them in the lower concrete portion of the dam, then bolted them to the masonry base for stability. Babcock also had the builders lay wire cables horizontally across the arch to fortify the structure. (SDCL.)

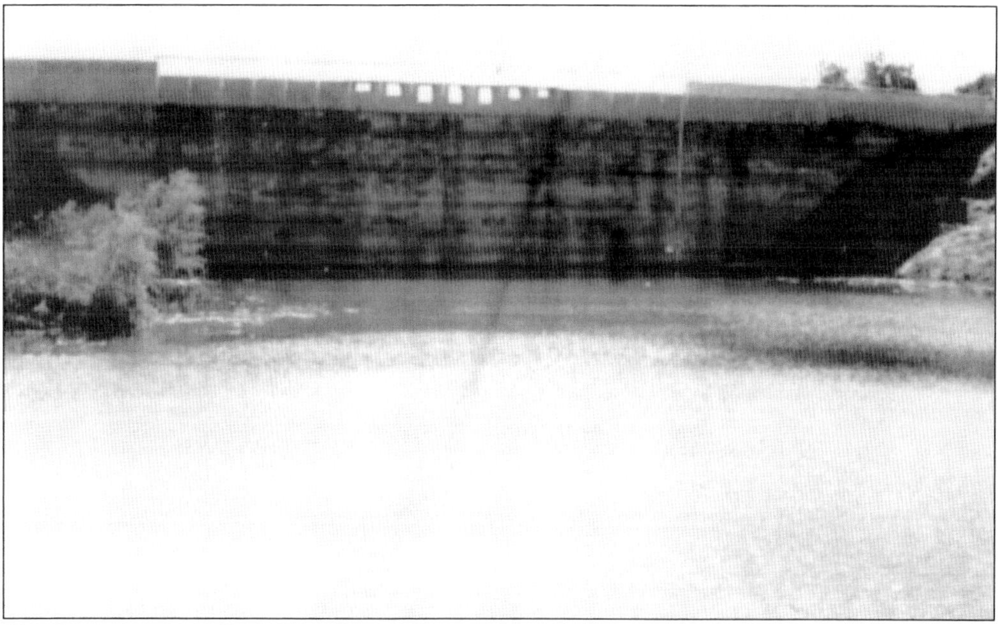

The dam's unique spillway was a set of seven gated notches sitting five feet below the crest in the center of the dam. The dam survived the flood of 1916, but it was debatable if the spillway system was the reason. For safety reasons, city engineers later removed the notches to create a simple open, overflow spillway that drops in two steps in the center of the structure. (SDCWHA.)

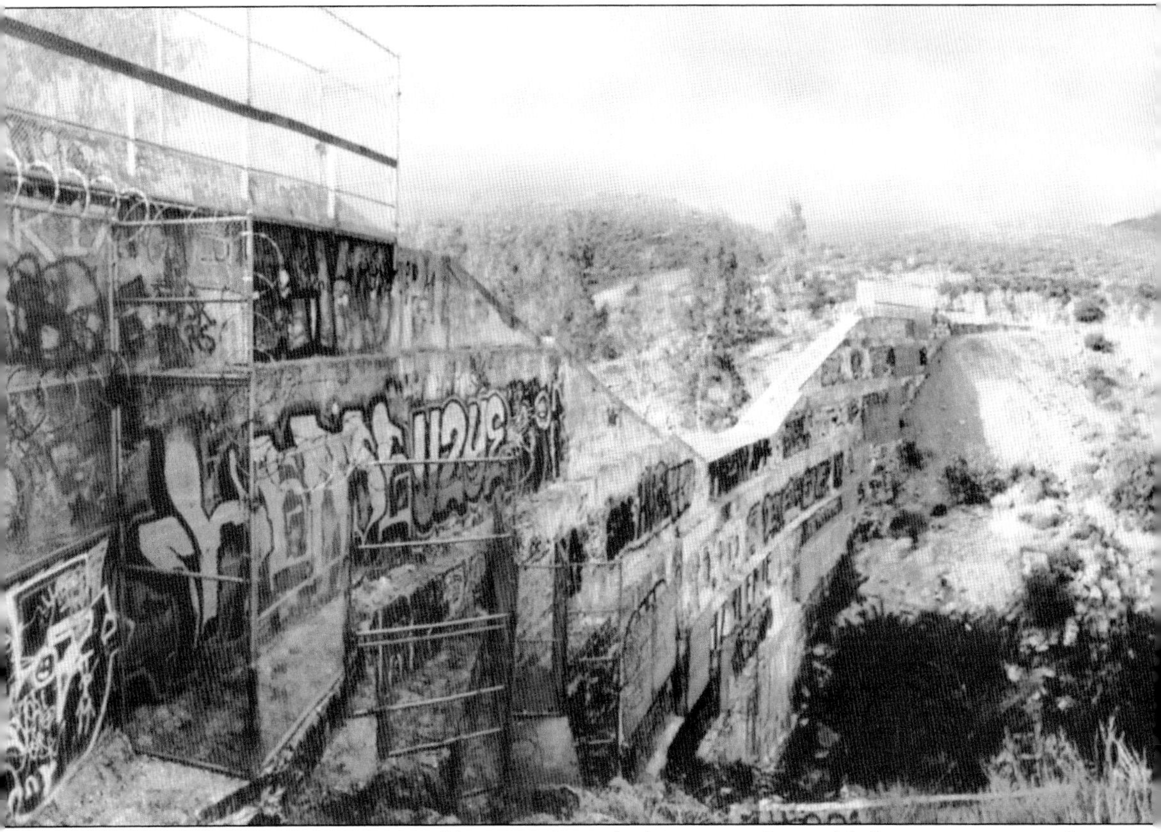

The Upper Otay Dam was an odd project from the beginning. It quickly became apparent that with the dam's location on a seasonal creek with a limited 12-square-mile catchment, any natural flow would leave the reservoir nearly empty except in flood years. Shortly after its completion, the SCMWC managers considered transporting water from the Cottonwood Creek catchment via an aqueduct and storing water in the reservoir, but they soon abandoned the idea. In the 1950s, the graffiti-covered dam and reservoir found its niche as a national hatching facility for a strain of Florida largemouth bass. The city was able to stock all its reservoirs with fish from Upper Otay. A state-funded program is also in the process of removing non-native plants, replanting native species, and improving the area as a recreational site. The space is now called Upper Otay Lake. (Author's collection.)

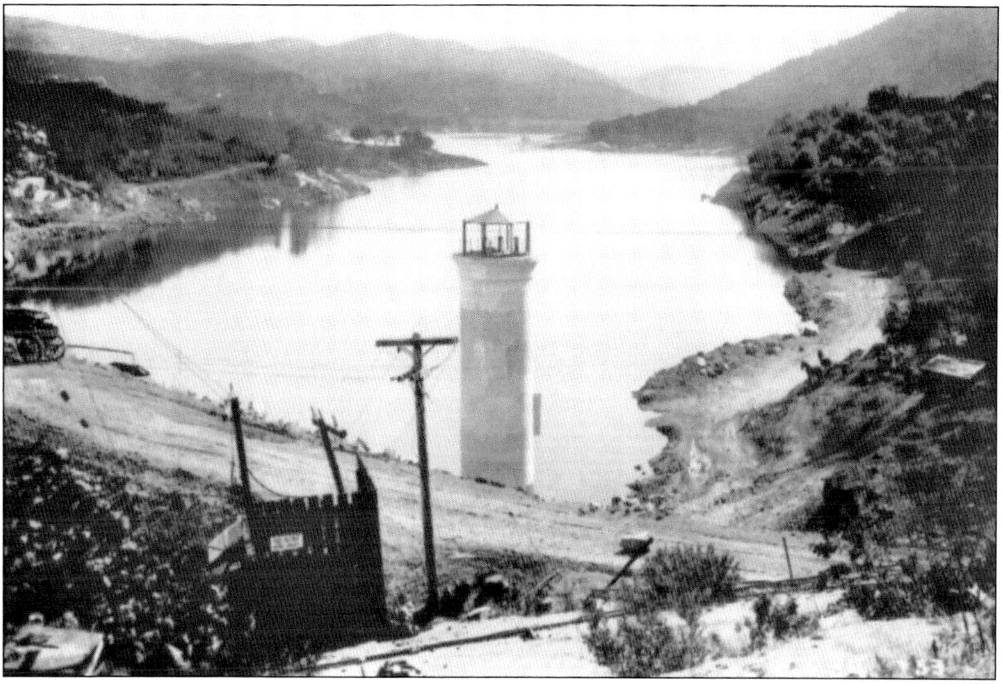

About the time Upper and Lower Otay were under way, the Escondido Irrigation District started construction of Wohlford Dam. Locals called the north county's first dam Bear Valley Dam, then the Escondido Dam, then eventually Wohlford Dam, after the original property owner, A.W. Wohlford. Sitting several miles north and above Escondido, the dam conserved the waters of the San Luis Rey River. A crew of 75 constructed the 76-foot-high, 397-foot-long earth-and-rock-filled dam. The builders lined the upstream face with rough and thick redwood planks placed vertically. With no true quarry available, they took the rock from the outcroppings on the sides of the surrounding canyons and hauled it by horse cart to the dam, where derricks placed the material. The dam cost the district $110,059. (Both, SDCWHA.)

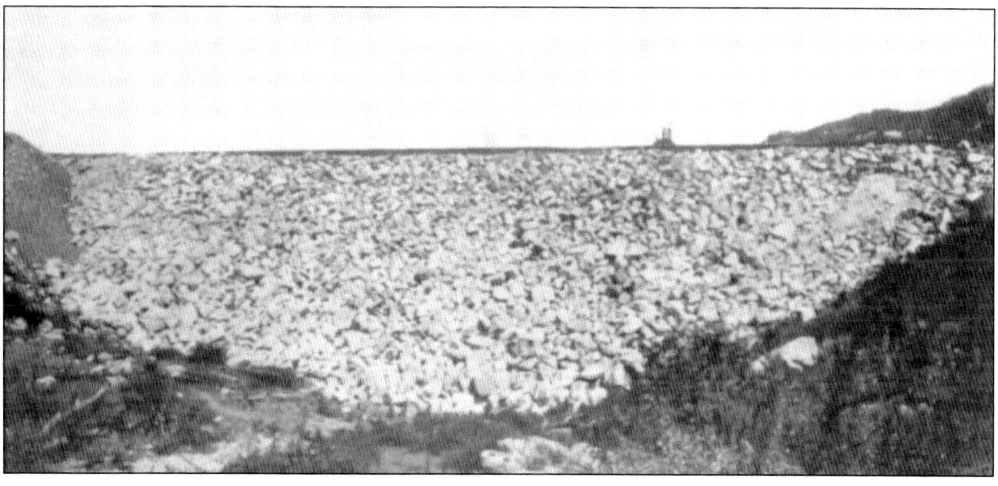

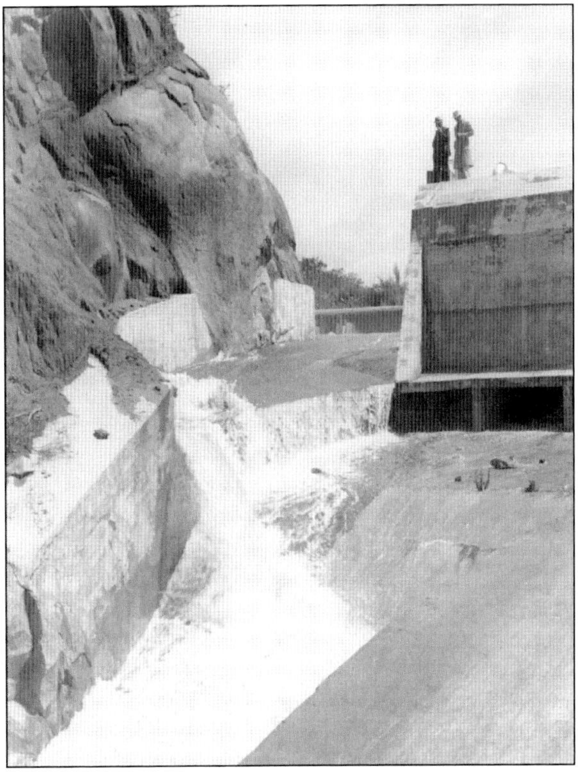

Wohlford Dam was the first rock-filled dam constructed in California for irrigation purposes. The original 25-foot-wide spillway featured a face of 6,000 yards of hand laid dry cobble and sat 70 feet above the streambed. An improved spillway featured flashboards on the crest to increase the storage capacity. Water was drawn from the reservoir through a 24-inch clay pipe set in concrete. (Courtesy of Vista Irrigation District.)

During Wohlford's construction, workers placed material on the dam with a trolley on a wire cable manipulated by hand with a windlass and rope. Sometimes they used an automobile to pull the line. The cable ran 600 feet above the riverbed and spanned 1,500 feet. The district planned to fill the reservoir with water transferred via an aqueduct. The project cost $116,328. (Courtesy of Vista Irrigation District.)

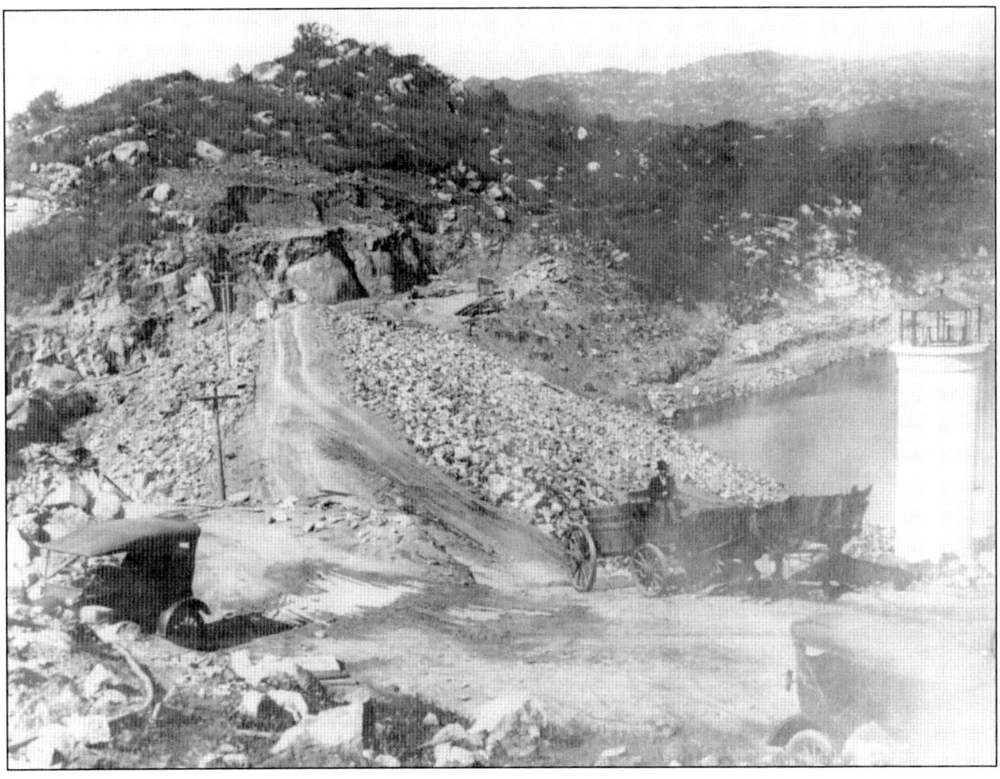

The flood of 1916 overflowed and damaged Wohlford, which necessitated repairing the facing and enlarging the spillway. In 1922, concern over future floods motivated the Escondido Mutual Water Company and the San Diego County Water Company to rebuild and enlarge the dam. In 1924, engineers used the hydraulic process to make the upstream face impervious and hopefully strengthen the dam. They also raised the overall height of the dam to 100 feet. Engineers also changed the existing spillway to a six-barrel siphon system capable of handling a lager outfall. The additional 24 feet enhanced the capacity, but in the opinion of many engineers, the construction methods left the structure with questionable stability during an earthquake. (Courtesy of Vista Irrigation District.)

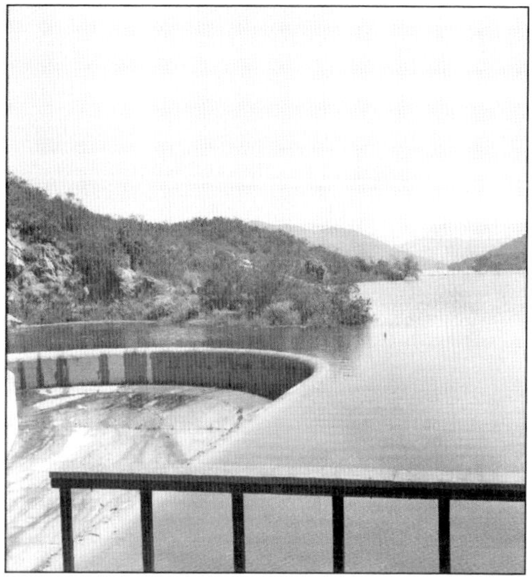

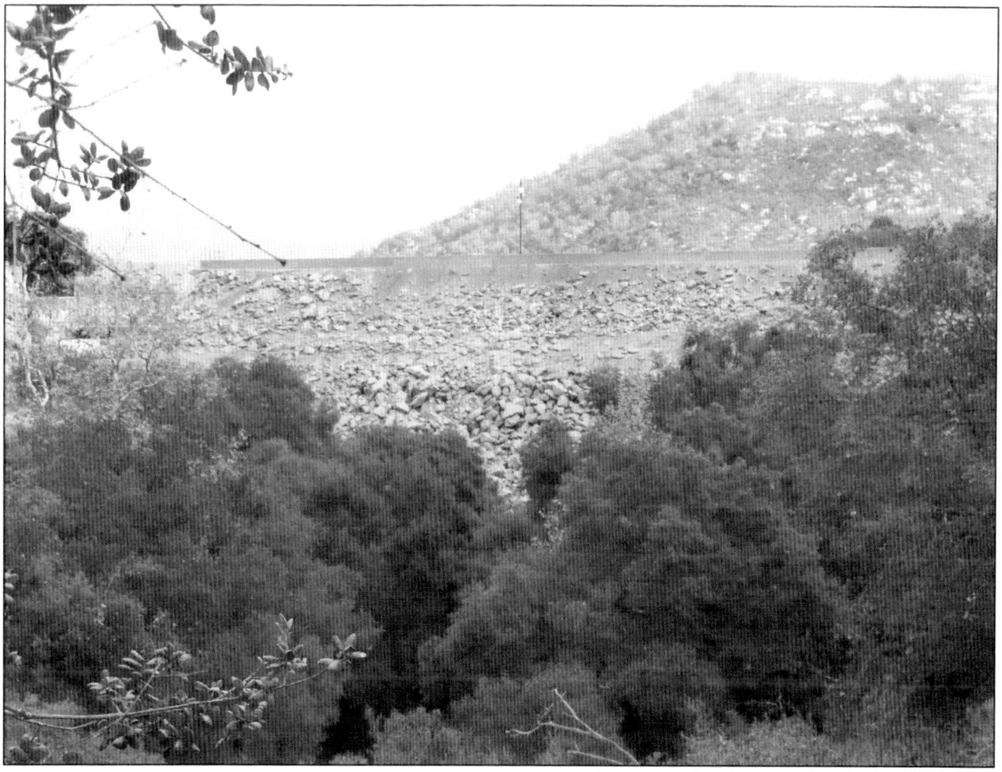

The undersized and leaky La Mesa Dam motivated the dam's new owner, the Cuyamaca Water Company, to replace it with a larger, safer dam in 1917. Company manager Ed Fletcher became intrigued with John S. Eastwood's innovative design concept for a multiple-arch dam because it offered the possibility of building a large storage dam with less material and in less time. Arched dams were essentially a series of cylindrical arches supported by buttresses and horizontal bracing between the buttresses, and Eastwood believed the application of arches and buttresses made his dams stronger and safer. The arched dam did require less concrete to build, but forming the arches required detailed carpentry work. (SDCL.)

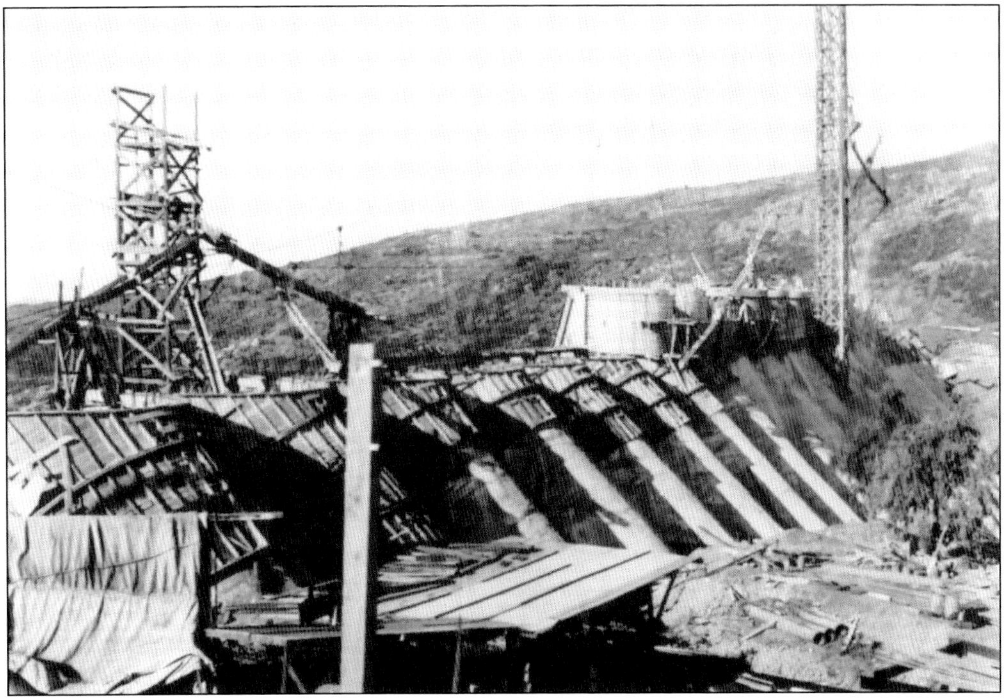

The new La Mesa Dam, shown at the dedication in 1918, was a reinforced concrete multiple-arch dam that stood 117 feet high and was almost 900 feet long. The dam cost $118,905 to construct. It featured 27 arches, each with a 30-foot span supported by solid buttresses, and had a siphon spillway on the west end. The dam was renamed for owner James Murray in 1924. (Both, SDCWHA.)

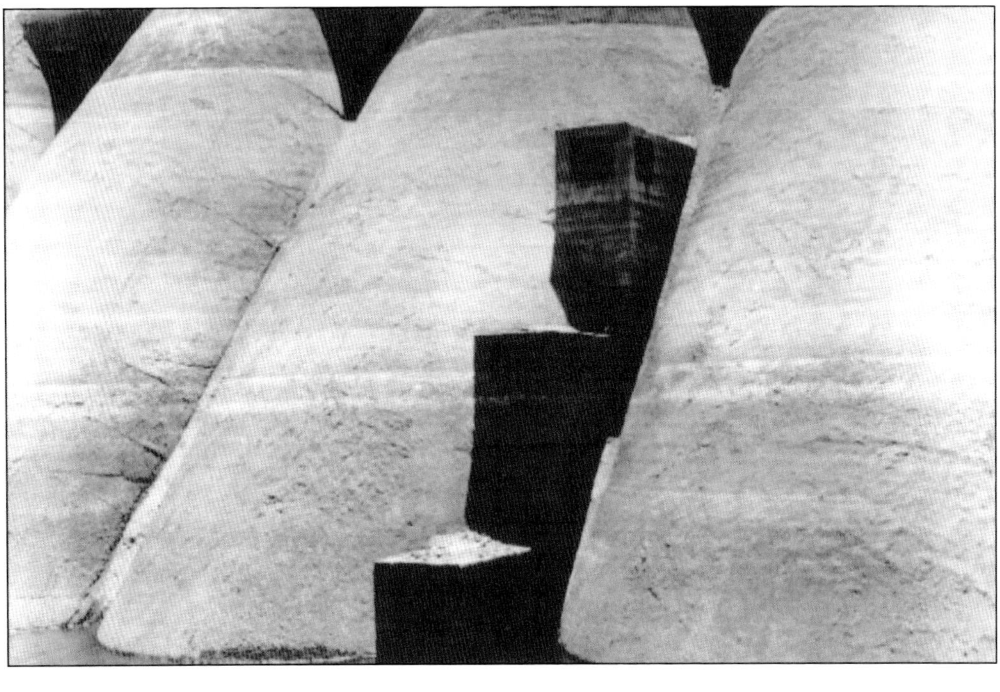

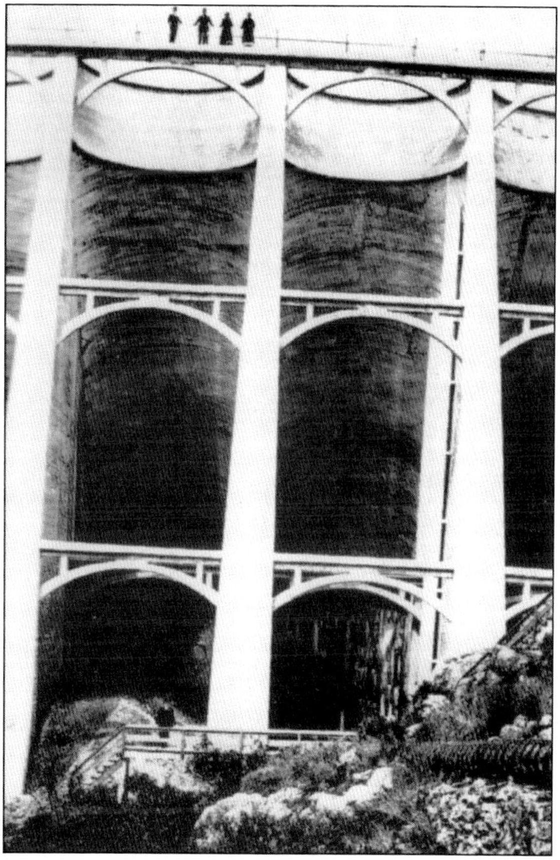

The Cuyamaca Water Company sold Murray Dam to the La Mesa, Lemon Grove, and Spring Valley Irrigation District in 1926. After years of realignment among area water districts, Murray became the property of the City of San Diego in 1961. The dam received a new outlet tower in 1943, and to this day, the reservoir continues as a storage facility for the city water supply system. (Both, SDCL.)

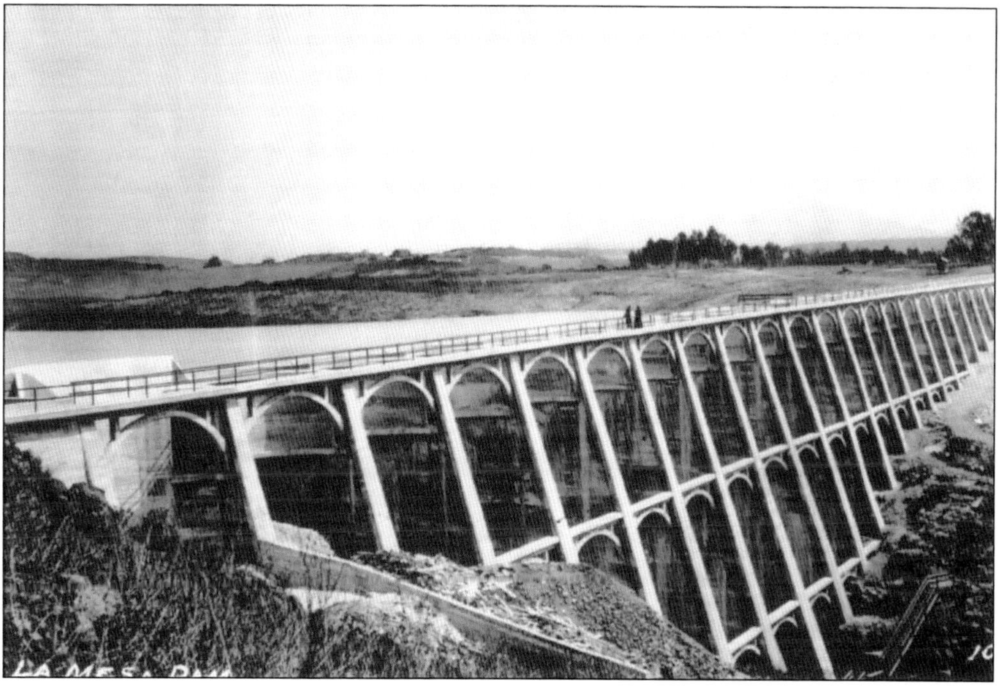

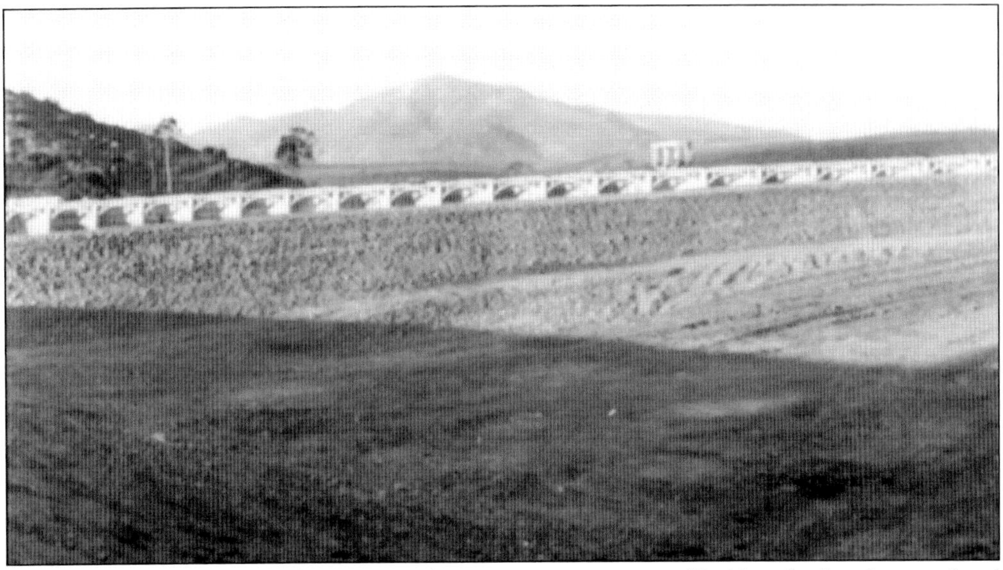

The city constructed Alvarado Dam in 1952 to serve as a storage and holding facility that regulated the downstream flow of water released from Murray Reservoir. Water from Alvarado Reservoir passed to the filtration plant, then into the city distribution system. Locally, the plant is noted for its ornamental Spanish Colonial bell tower. (SDCWHA.)

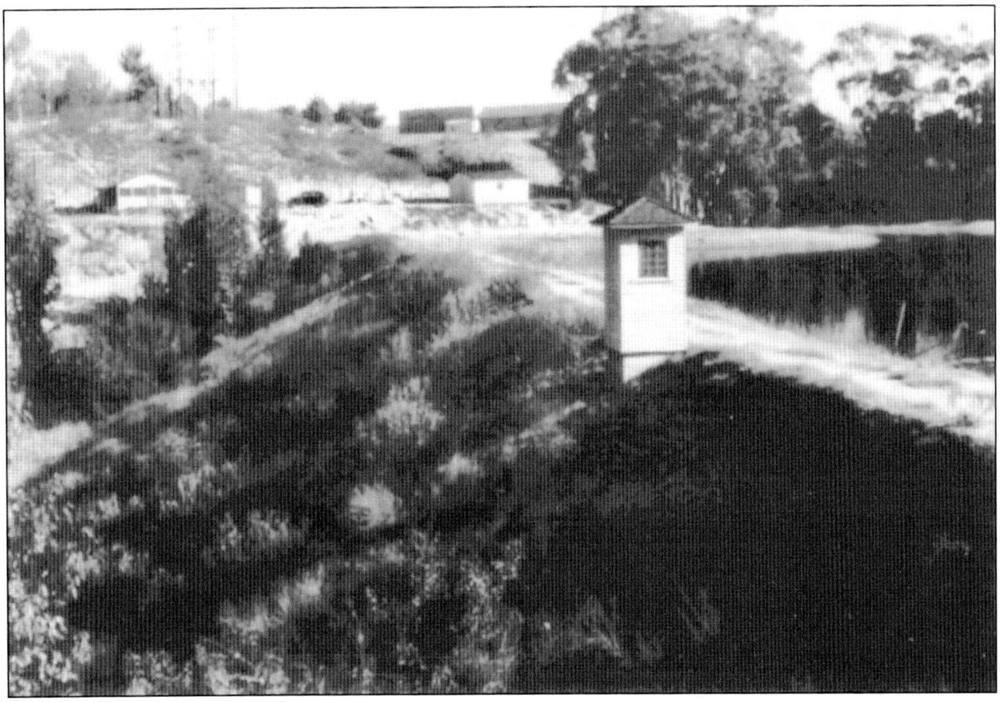

Engineers Hiram N. Savage and E.F. Tabor supervised the construction of the earth-filled Chollas Heights Dam for the SCMWC in 1901. The dam is located about four miles east of San Diego. The water company intended the Chollas Heights Reservoir to be the terminal point for the 19-mile Bonita Pipeline, which brought water from the Lower Otay Reservoir. (SDCWHA.)

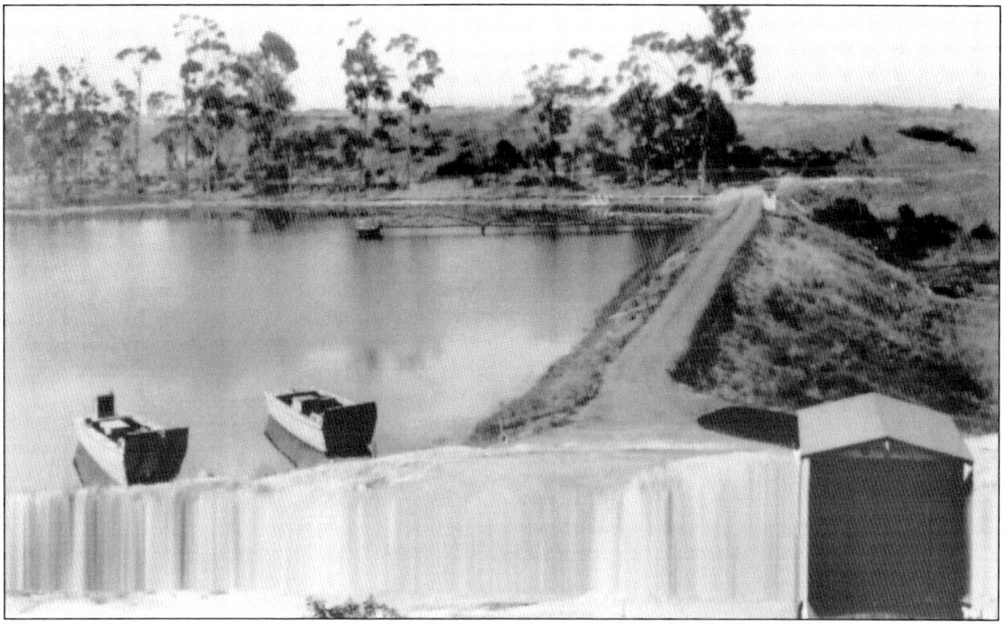

The Chollas Dam stands 56 feet high and 526 feet long, and like the original Lower Otay, it features a core of steel plate with reinforced concrete covered with earth and rock. The reservoir receives no runoff from a watershed, hence it is a storage facility. It has 17 acres of surface area. (SDCWHA.)

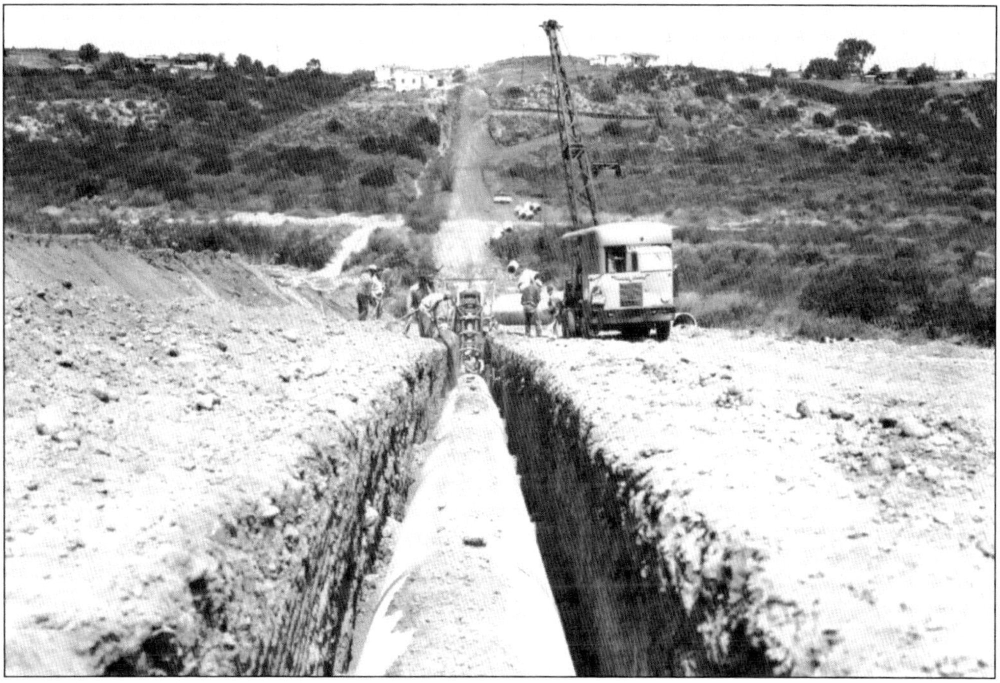

The water stored in Chollas was moved to University Heights via four miles of iron pipes to a filtration plant, and then distributed to the city. The city started construction of the 40-inch redwood-stave pipeline in 1900, but with constant interruptions, the project was not completed until 1906. (SDCWHA.)

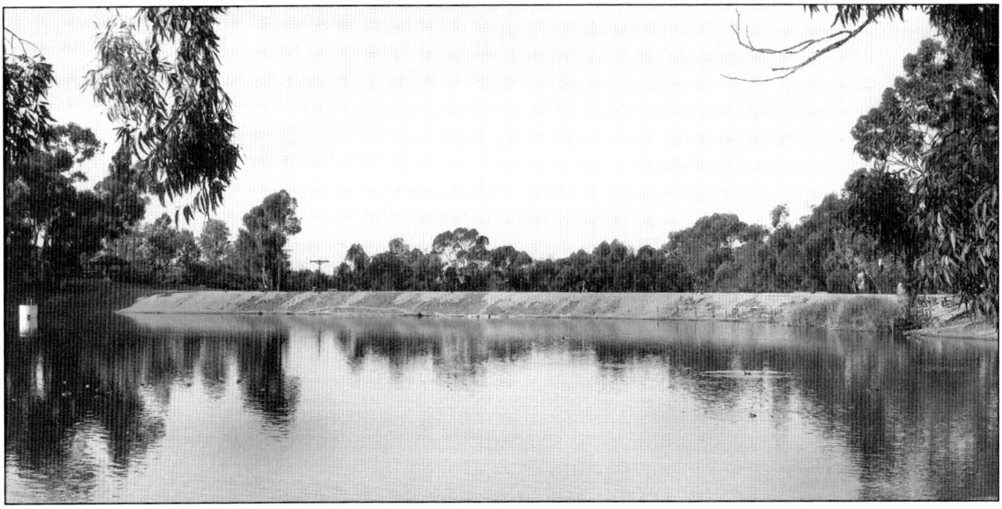

The city decommissioned the water treatment plant at Chollas in 1950. The San Diego recreation department took control of the space in 1966 and redesigned it into a fishing lake for anyone under 15 years of age. Tall eucalyptus trees surround Chollas Lake Park, which features exercise areas and a popular jogging loop for the residents of Lemon Grove, La Mesa, and Spring Valley. (Author's collection.)

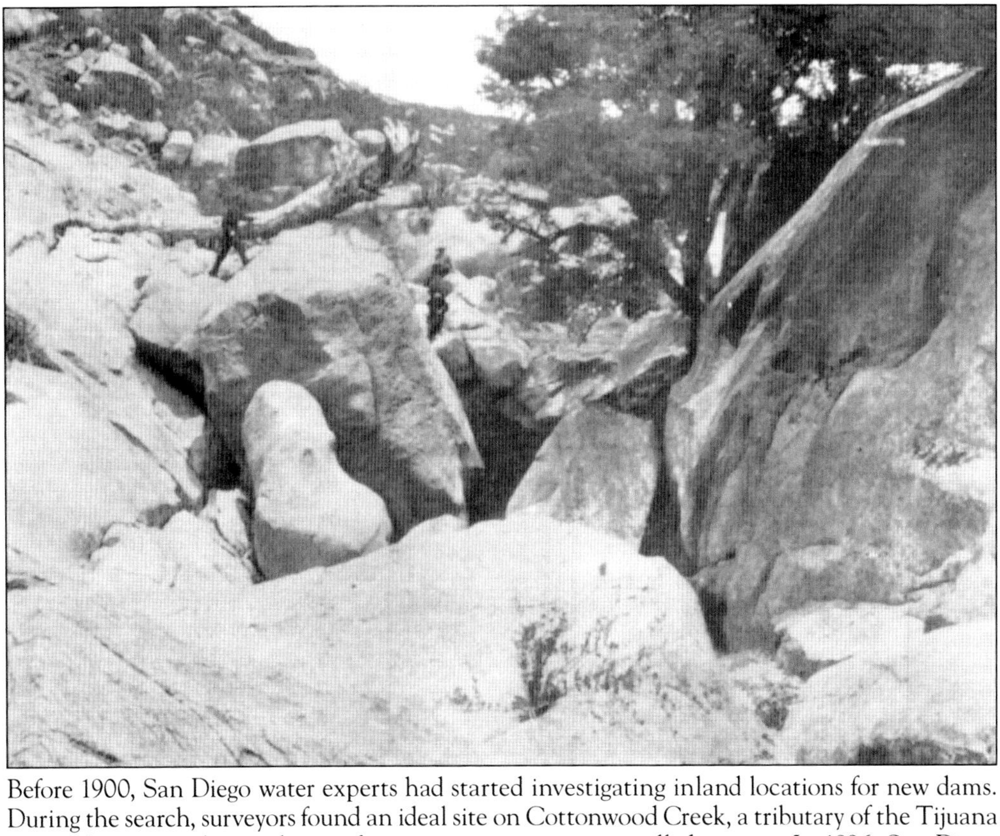

Before 1900, San Diego water experts had started investigating inland locations for new dams. During the search, surveyors found an ideal site on Cottonwood Creek, a tributary of the Tijuana River, about 40 miles southeast of town in a narrow, steep-walled canyon. In 1896, San Diego voters approved $1.5 million in bonds, and the Morena Dam project became reality. (SDCL.)

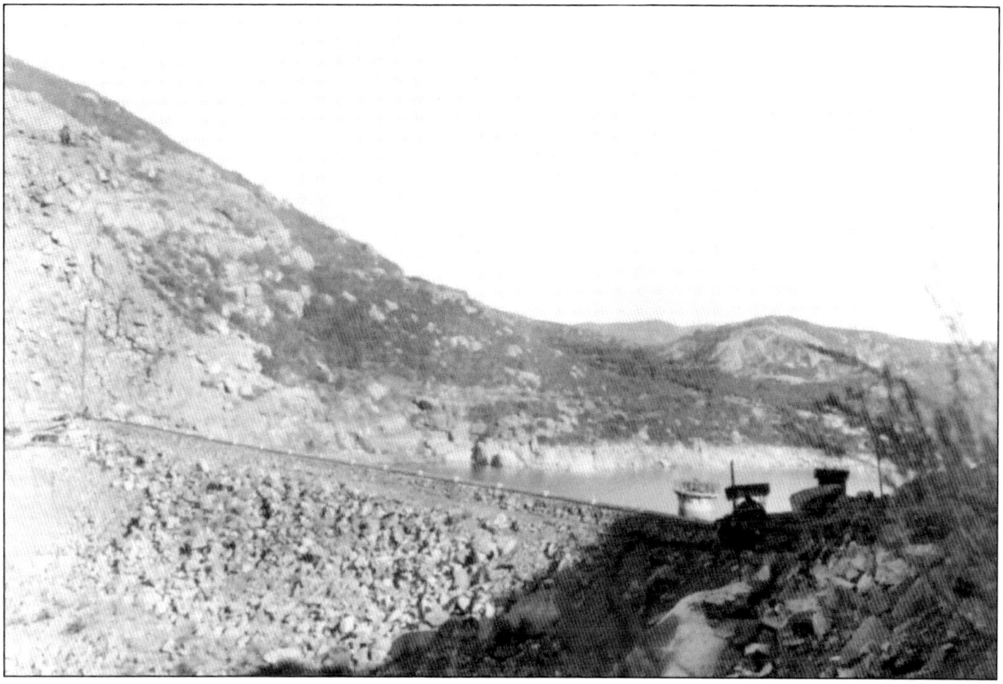

The Mount Tecate Land and Water Company started work at Morena in 1895, but discontinued operations in 1897. Twelve years later, the SCMWC bought the company and resumed work. Elisha S. Babcock headed the project, and as at Otay, the job was an amateur undertaking. *Harper's* magazine called the dam one of the wonders of the world, but locals called it the "gunny sack" dam after laborers stuffed sacks into the leaky cracks forming on the face of the dam. City engineer Edwin Capps made a visit to the site to view the dam and charged the builders with extraordinary negligence. Leaks, a lack of funds, and poor engineering doomed the original effort, and forced the SCMWC to suspend operations in 1898. (Both, SDCL.)

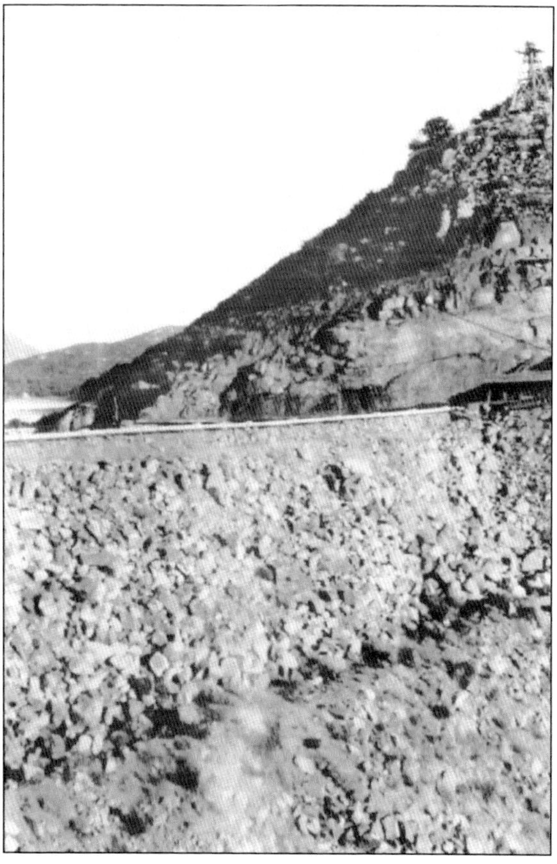

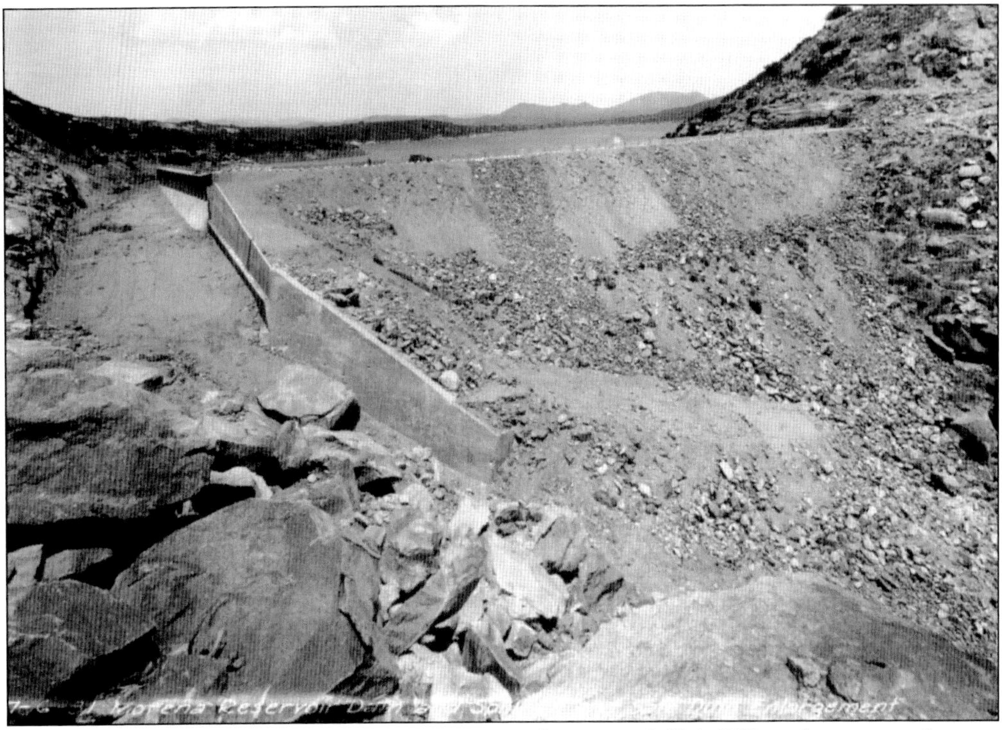

Owner John Spreckels brought in San Francisco civil engineer M.M. O'Shaughnessy to take over the Morena project in 1909. Concerned about the dam's stability, O'Shaughnessy ordered a new foundation for the base consisting of 50-foot-thick, 7-foot high granite block secured to the lower slope and had the upper slope of the dam face covered with reinforced concrete slabs. (SDCL.)

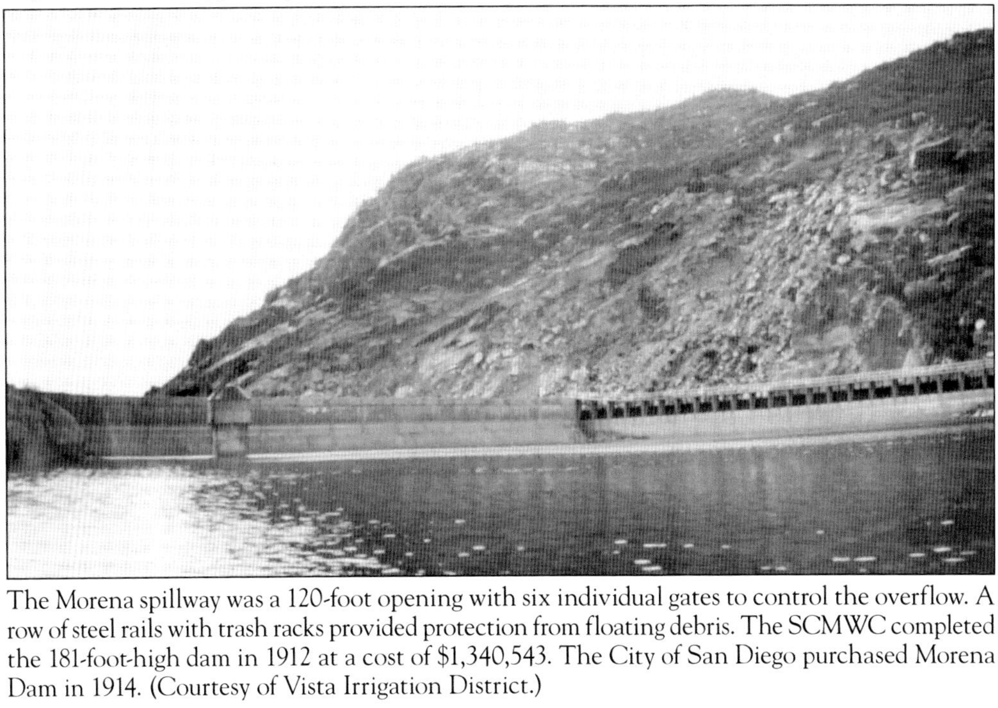

The Morena spillway was a 120-foot opening with six individual gates to control the overflow. A row of steel rails with trash racks provided protection from floating debris. The SCMWC completed the 181-foot-high dam in 1912 at a cost of $1,340,543. The City of San Diego purchased Morena Dam in 1914. (Courtesy of Vista Irrigation District.)

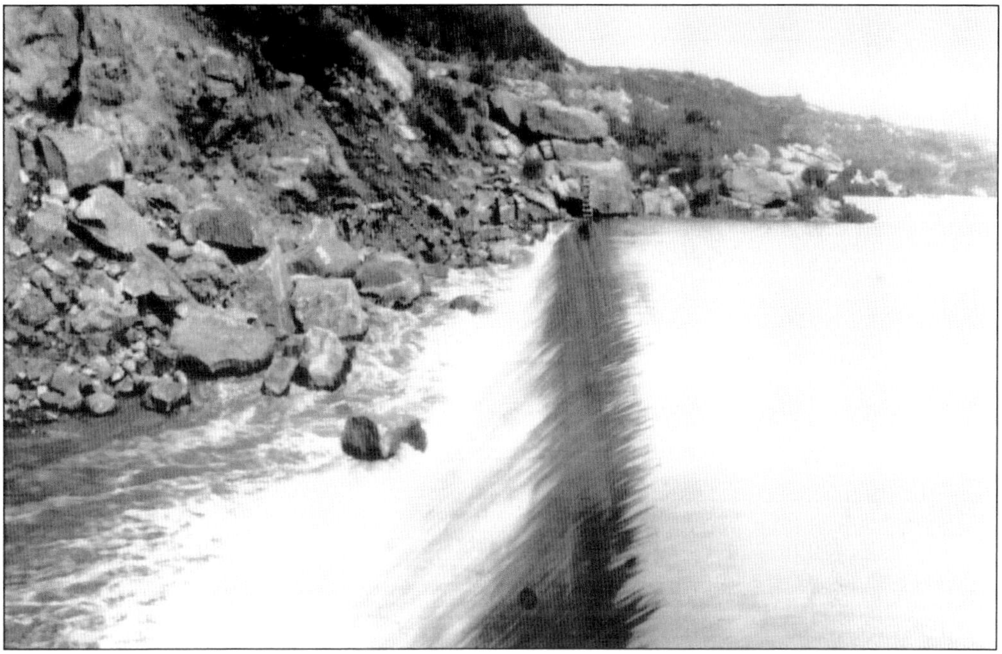

The flooding of 1916 that destroyed Lower Otay Dam and partially ruined Sweetwater Dam also nearly overtopped Morena Dam. The trash racks engineers designed for protecting the spillway from flotsam became so clogged with debris that the reservoir waters were three feet above the spillway. The dam survived, but the near disaster prompted the city to raise the dam's vertical face another five feet in 1917. The city increased the height of the dam an additional 10 feet in 1922–1923, then four more feet in 1930. This work increased the capacity from the original 48,000 acre-feet to 58,000. (Both, SDCWHA.)

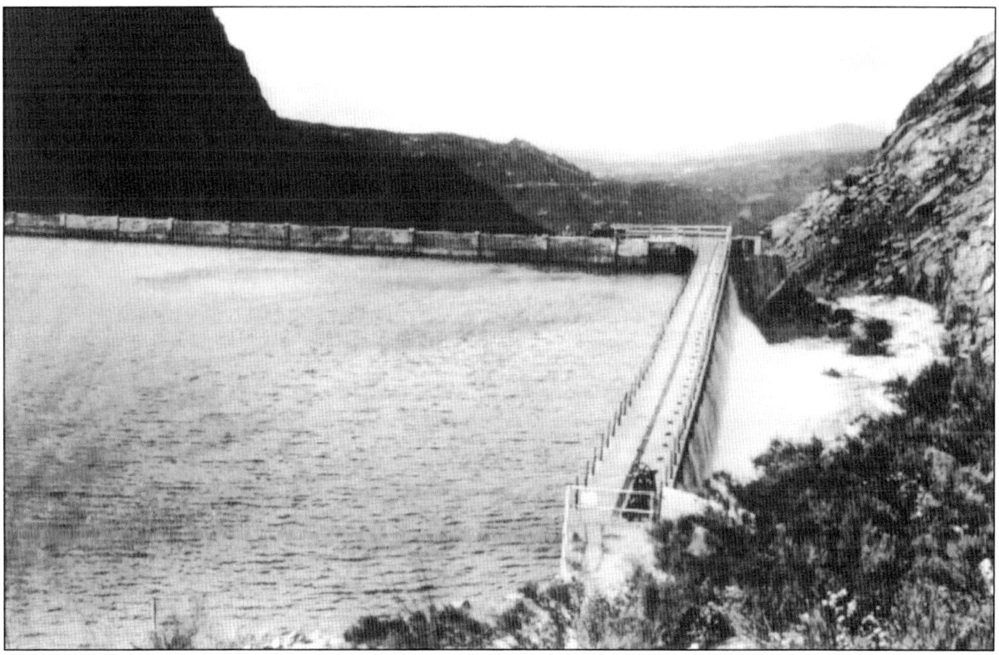

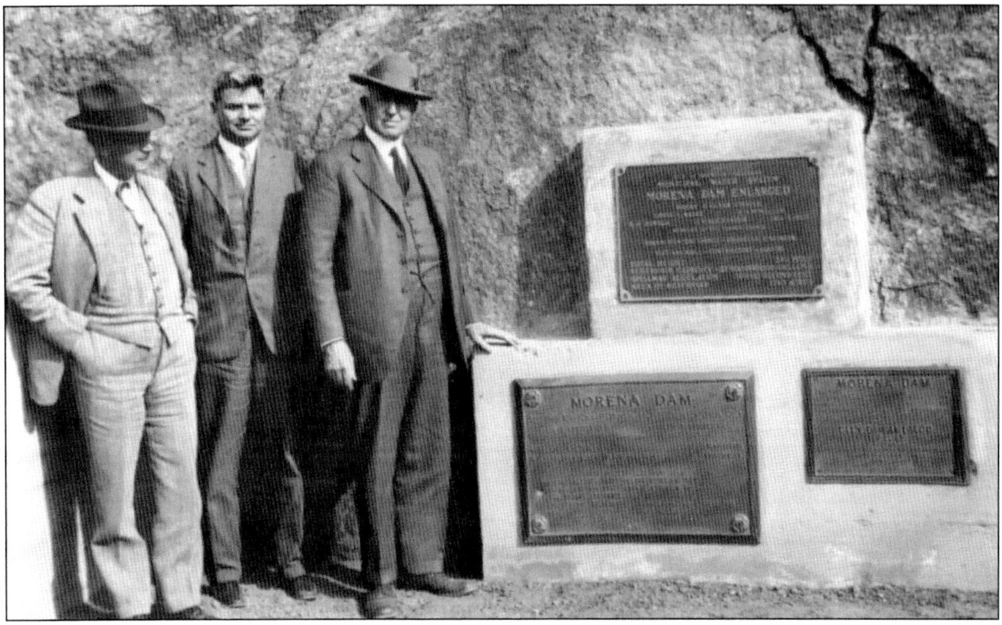

City engineer Hiram Savage and city officials rededicated Morena Dam in 1931 after the completion of the extensive renovations. The spillway had been improved, and the dam was heightened to 273 feet to increase the reservoir storage capacity. The dam has a graceful 150-foot contour and a crest thickness of 16 feet, which engineers later topped with a three-foot wave coping. Morena is the most remote of all the city dams and sits at the highest elevation of any dam in the county at 3,000 feet. Savage designed Morena Dam to conserve water during storm events and capture natural runoff to fill the reservoir. (Both, SDCWHA.)

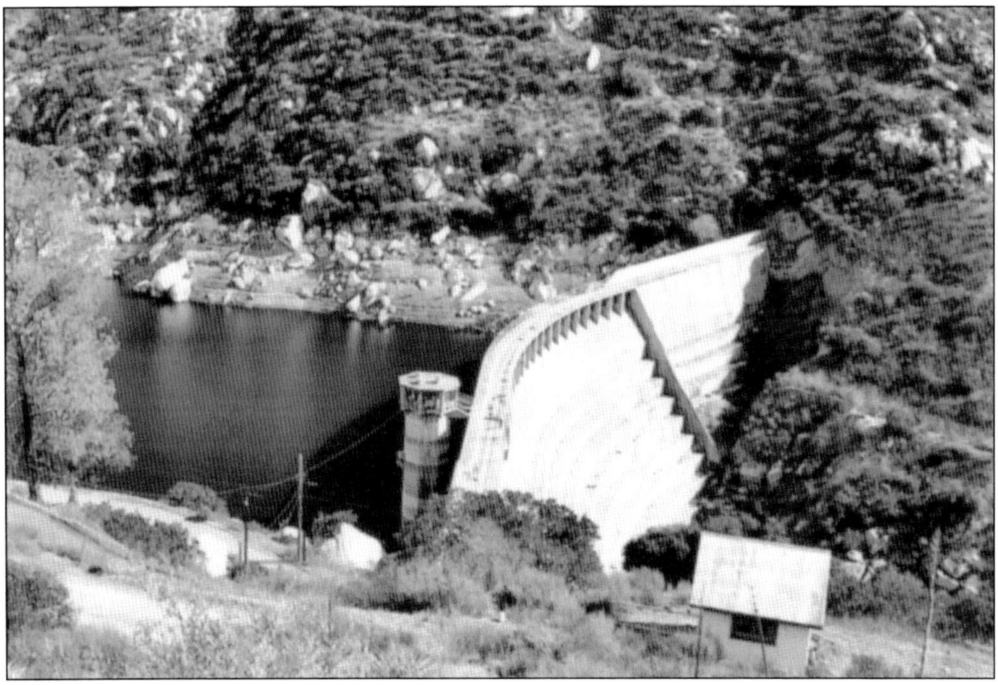

When San Diego businessman John D. Spreckels secured the contract to supply water to the city of San Diego from his water company in August 1906, he had resumed construction on the Morena Dam project. However, he had not yet solved the problem of how the company could move water from Morena to Lower Otay, where it could enter the city system. Spreckels decided an overland aqueduct was the logical solution. (SDCWHA.)

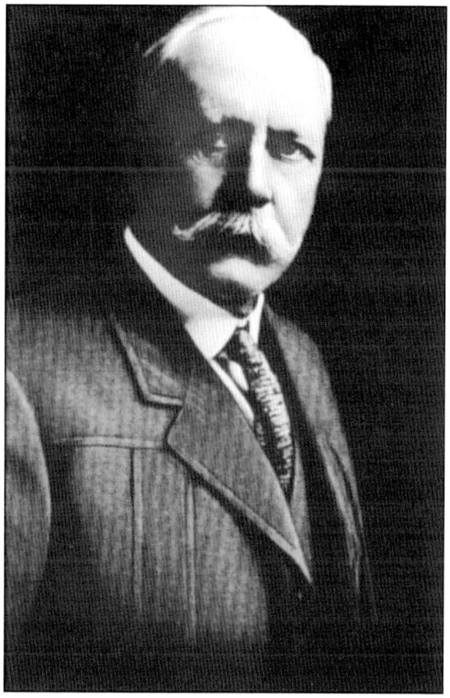

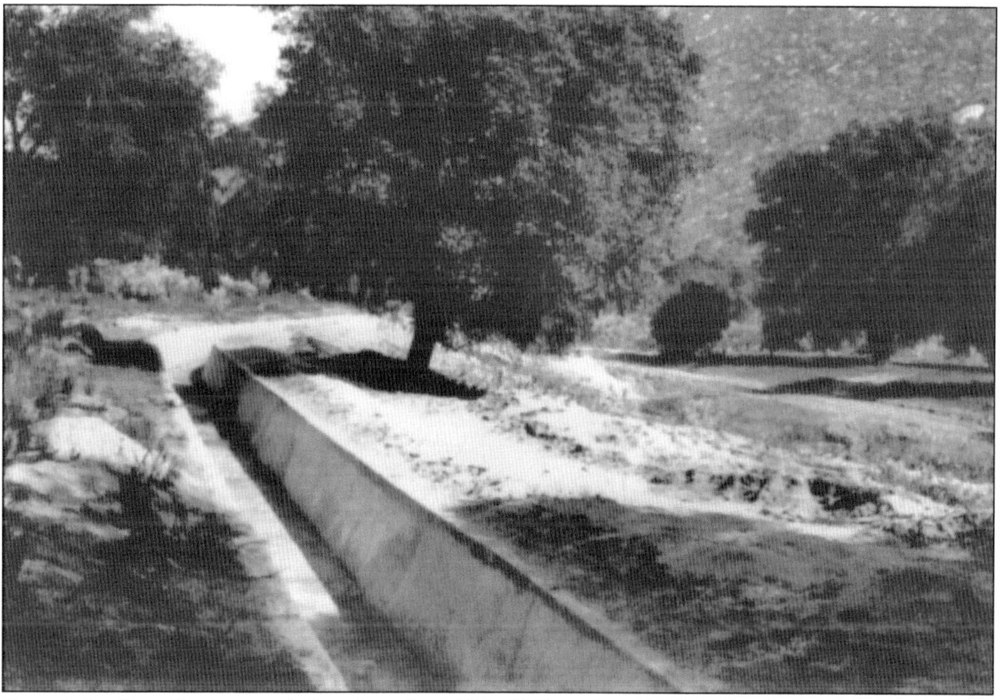

Spreckels allegedly designed the plans for M.M. O'Shaughnessy, his chief engineer, who started work on the conduit in August 1907. Working by hand in the steep, rugged terrain, the laborers created dirt channels, built open redwood flumes and concrete conduits, and excavated miles of tunnels to eventually connect Morena and Lower Otay. O'Shaughnessy completed construction on the 13.38-mile gravity-flow aqueduct in January 1909. (SDCWHA.)

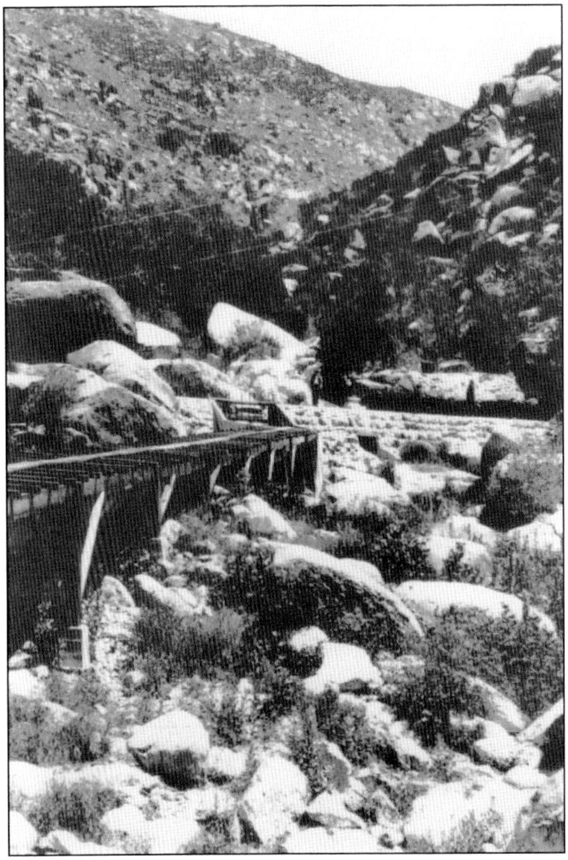

The conduit received water from intakes two miles above the future Barrett Dam, one on Pine Creek, which the dam later submerged, and one on Cottonwood Creek. The conduit received its name from its termination point at the head of Dulzura Creek and its proximity to the small town of Dulzura. The City of San Diego purchased the conduit along with Barrett Dam, Lower Otay, and the Chollas Heights Reservoir from the SCMWC in 1912 for $100,000. As the conduit deteriorated over time, the city made repairs and improvements and kept the original system in operation into the 1980s. In the 1990s, the city used helicopters to renovate the line, replacing the open sections with steel pipelines supported on concrete and steel standards and reconditioning the tunnels. The Dulzura Conduit remains in use today. (Both, SDCWHA.)

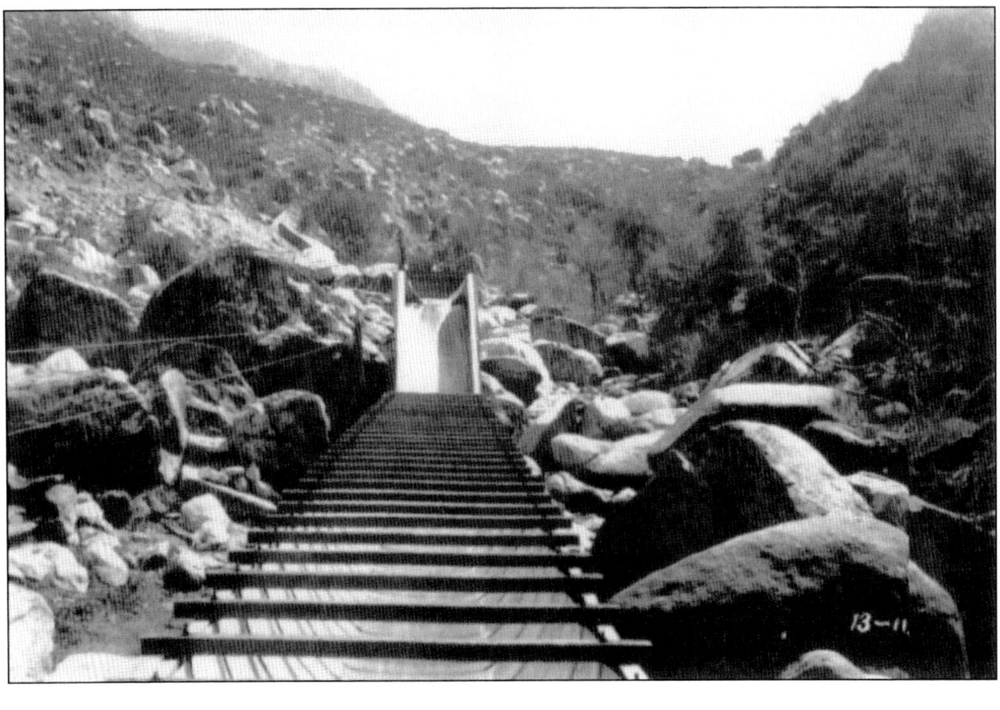

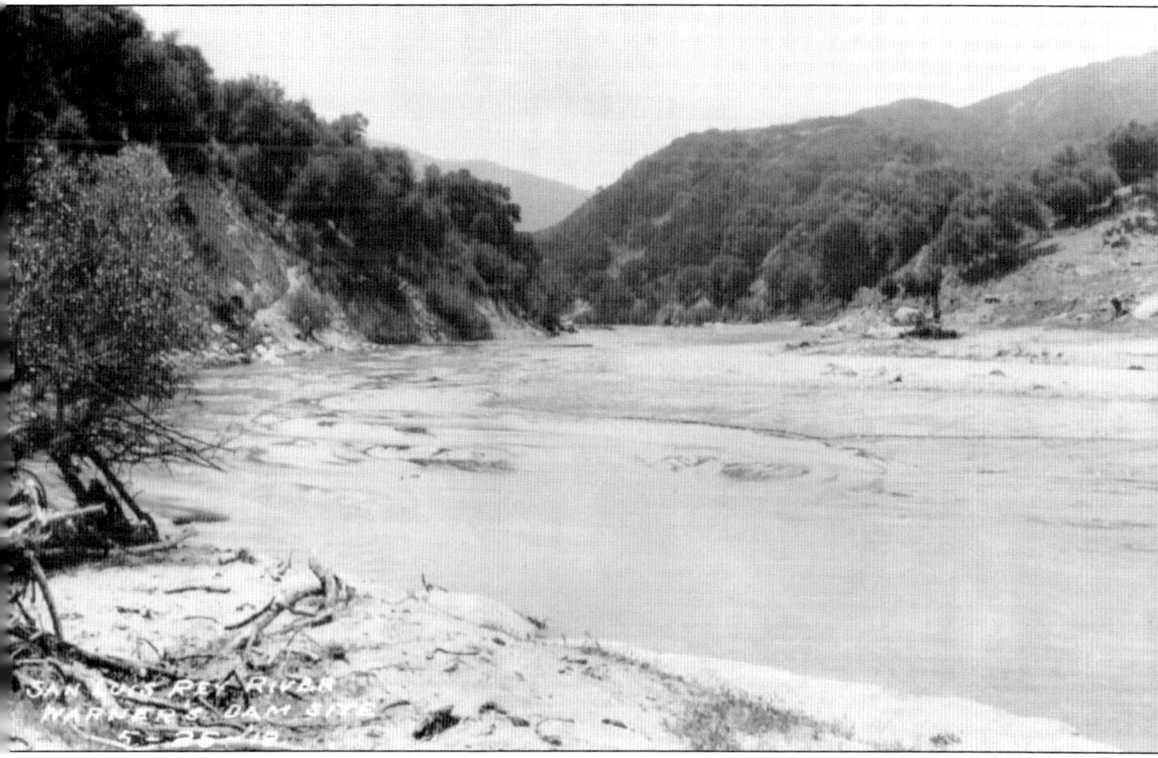

Just after the end of the 19th century, U.S. Grant Jr. and George Puterbaugh, representing the Warner family interests, proposed to build a dam in the far north county at the head of the 70-mile-long San Luis Rey River. The river had its headwaters in the Palomar and Santa Rosa Mountains and looped around the outer edges of the north county before it flowed into the Pacific near the town of Oceanside and drained almost 560 square miles. But E.S. Babcock and his SCMWC mounted a spirited opposition, and the plan failed to materialize. In 1905, the Warner interests sold to Henry Huntington and his Pacific Light and Power Company. He believed he could built a dam and sell power to San Diego and Los Angeles. In 1911, Huntington in turn sold out to James Henshaw, who had acquired the majority of the water rights from the dam site to the coast. (SDCL.)

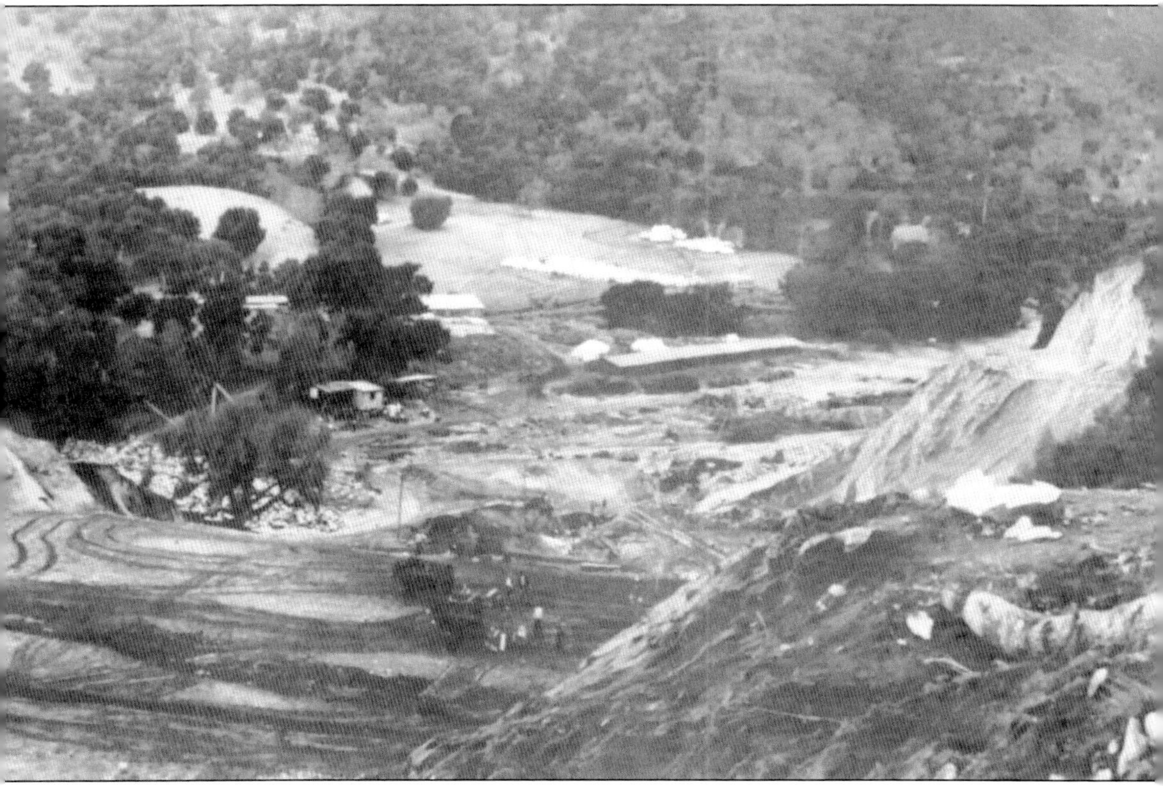

Ed Fletcher's Volcan Land and Water Company opened construction on the dam for Henshaw in 1914 but halted work when the City of San Diego investigated purchasing the system in 1915. The city opted out of the sale, and the San Diego County Water Company purchased the site and resumed work in June 1922 with engineers J.B. Lippincott and Rex Starr supervising the job. The threat of winter flooding pushed the builders to implement a 24-hour work schedule with the hope of completion by the first of the next year. Reportedly, 400 men and 200 mules were on the job. A geologic survey revealed that the site lacked the appropriate bedrock for a masonry dam, so the owners decided on a hydraulic fill dam. The hydraulic pool is seen in the core of the dam in this 1921 photograph. (Courtesy of Vista Irrigation District.)

Workers excavated the Henshaw site with mule teams, then covered the face of the 117-foot-by-660-foot dam with earth rock rip-rap from a hillside quarry. The Henshaw workers' construction camp had a dining room, a pool hall, a guesthouse, and a public school for the 25 camp children. James Henshaw and his family are shown visiting the $900,000 dam near completion in 1923. (Courtesy of Vista Irrigation District.)

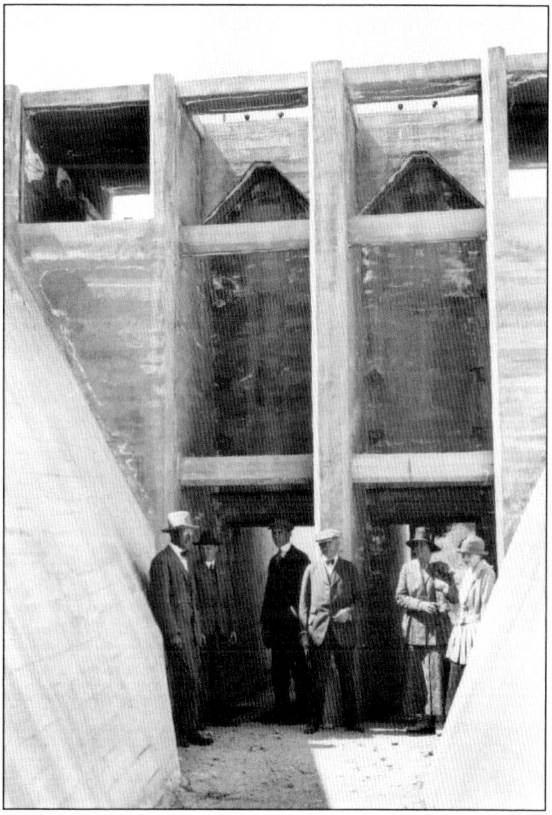

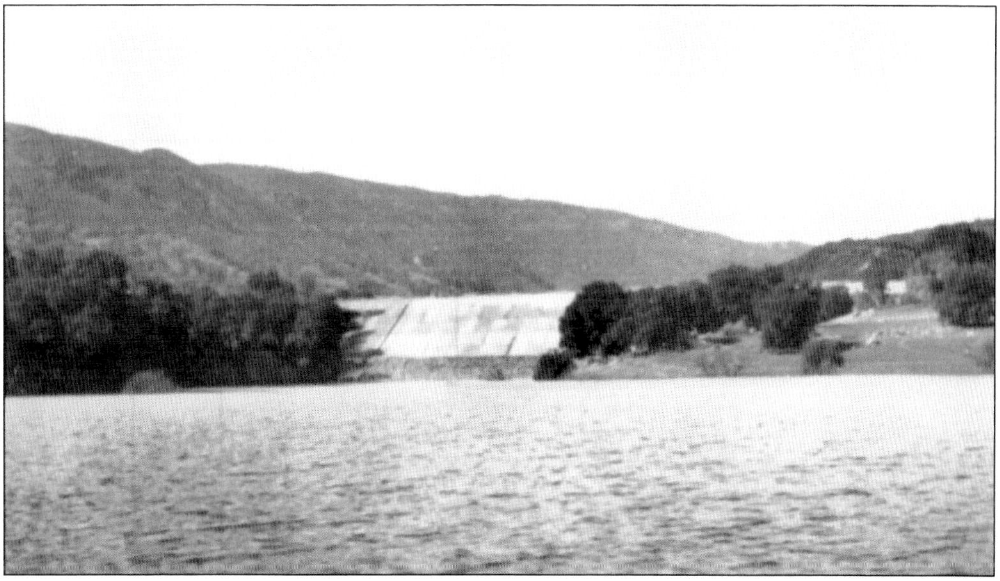

Workers faced Henshaw's core with 15-foot square panels of four-inch-thick reinforced concrete. There were two auxiliary dams on the left side of the main structure also faced with the concrete panels. The spillway was a 200-foot-long uncontrolled curved overflow. The dam was raised 10 feet in 1927 and a new spillway added. The Vista Irrigation District purchased the entire system in 1946. (Courtesy of Vista Irrigation District.)

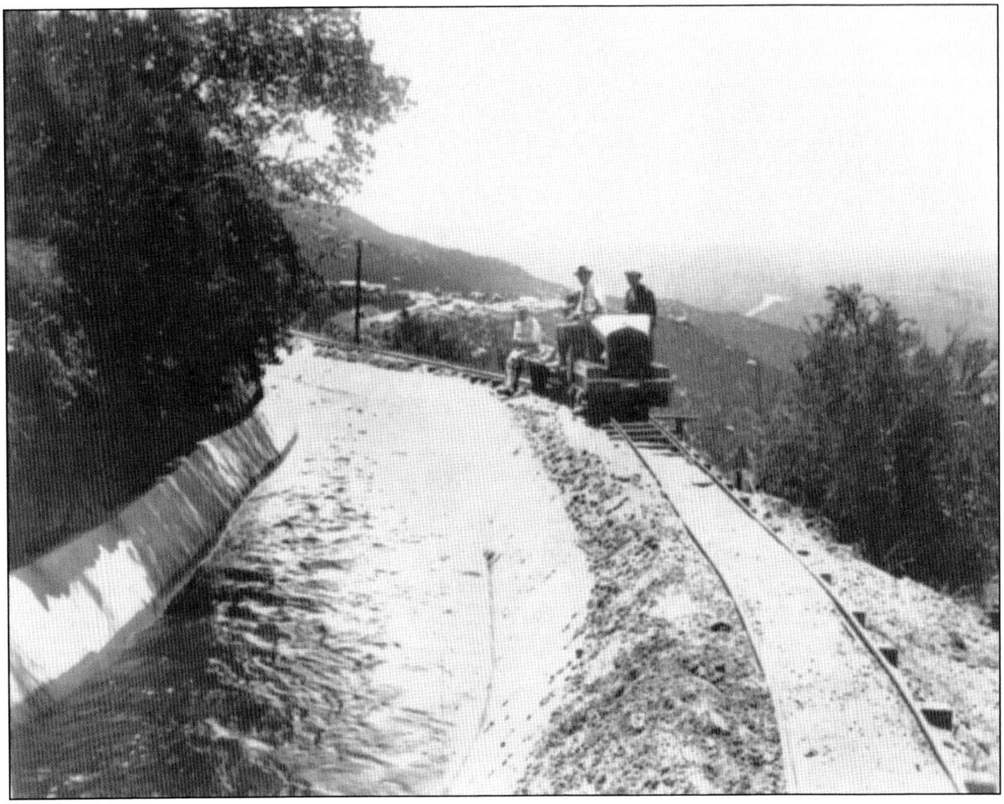

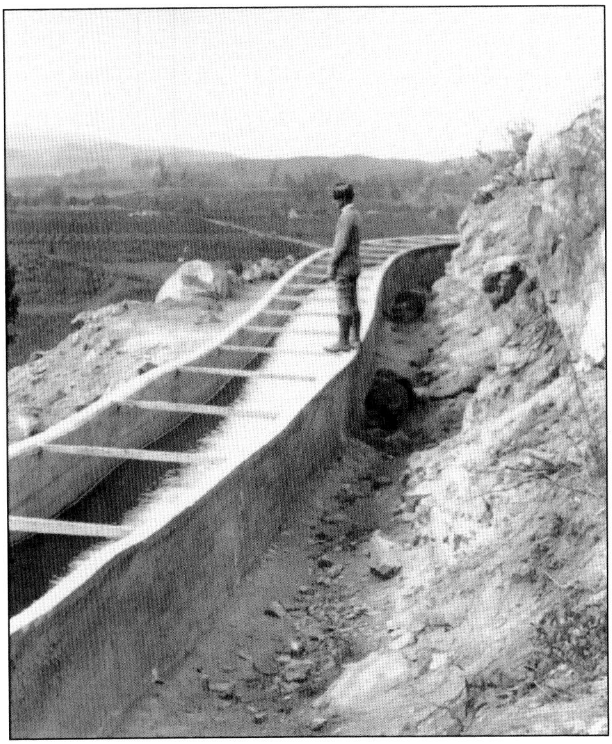

After the Vista Irrigation District acquired Henshaw, its engineers established an entire conveyance system to move the water from the reservoir to the town of Vista. Water from Henshaw Reservoir fed the six-mile Escondido Canal, which carried the water via a diverting dam to Lake Wohlford. The Vista Flume then transported the water from Wohlford to the Pechstein Reservoir, where it was moved into the treatment plant and distribution system. The system has an overall drop in elevation from 2,690 feet at Henshaw to 980 feet at the terminal reservoir in Vista, a descent of some 1,700 feet. (Both, courtesy of Vista Irrigation District.)

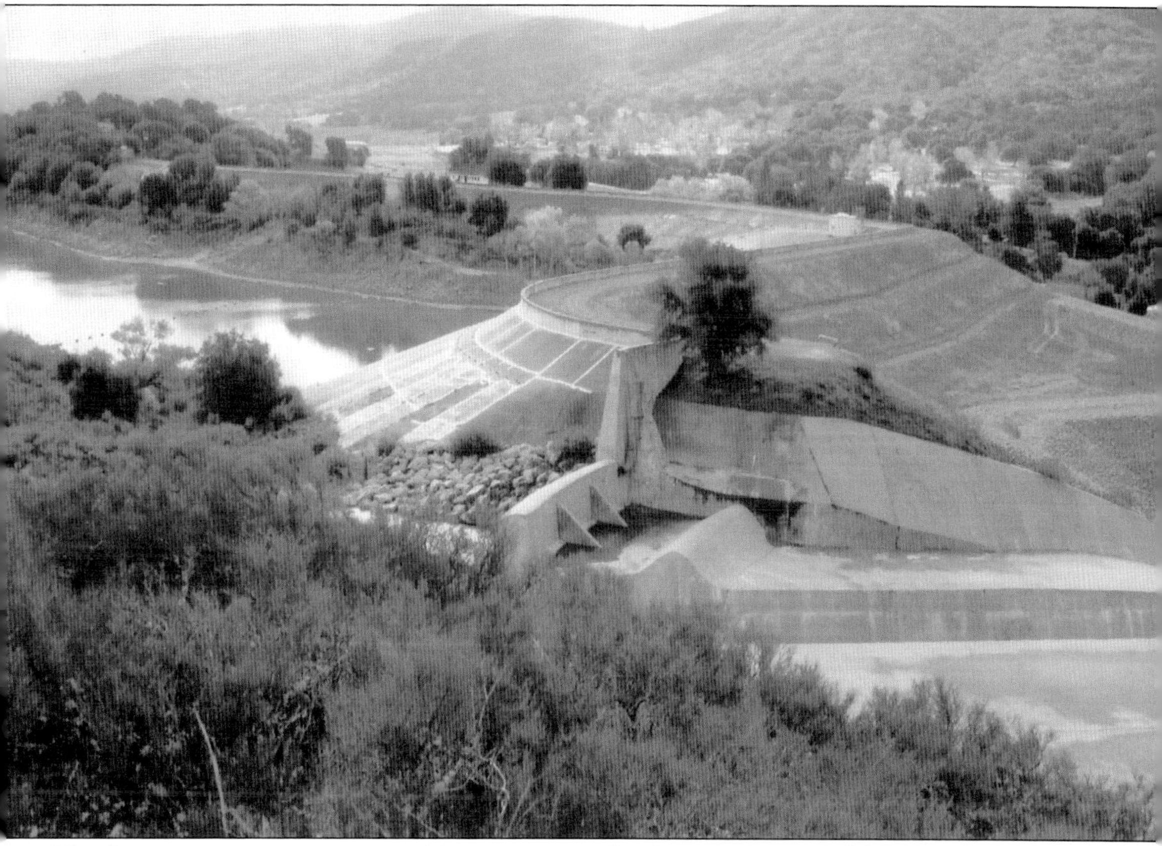

Henshaw never met expectations of the irrigationists but became the core of the Vista domestic system. The reservoir site for the dam was the bed of an old natural lake at the confluence of Carrizo Creek, Buena Vista Creek, and the San Luis Rey River. The reservoir, then the largest in the county, has never completely filled. The closest it came was following the floods in 1943. The reservoir was actually dry in 1951. Controversy arose in 1969 when five bands of north county Indians brought suit against the Escondido and Vista water agencies. In the 1880s, the Department of the Interior signed away its rights to the water passing through its reservation lands and granted James Henshaw a contract to build the dam. The Indians sued to recover the water but lost. In 2014, the parties finally reached a settlement. (Courtesy of Vista Irrigation District.)

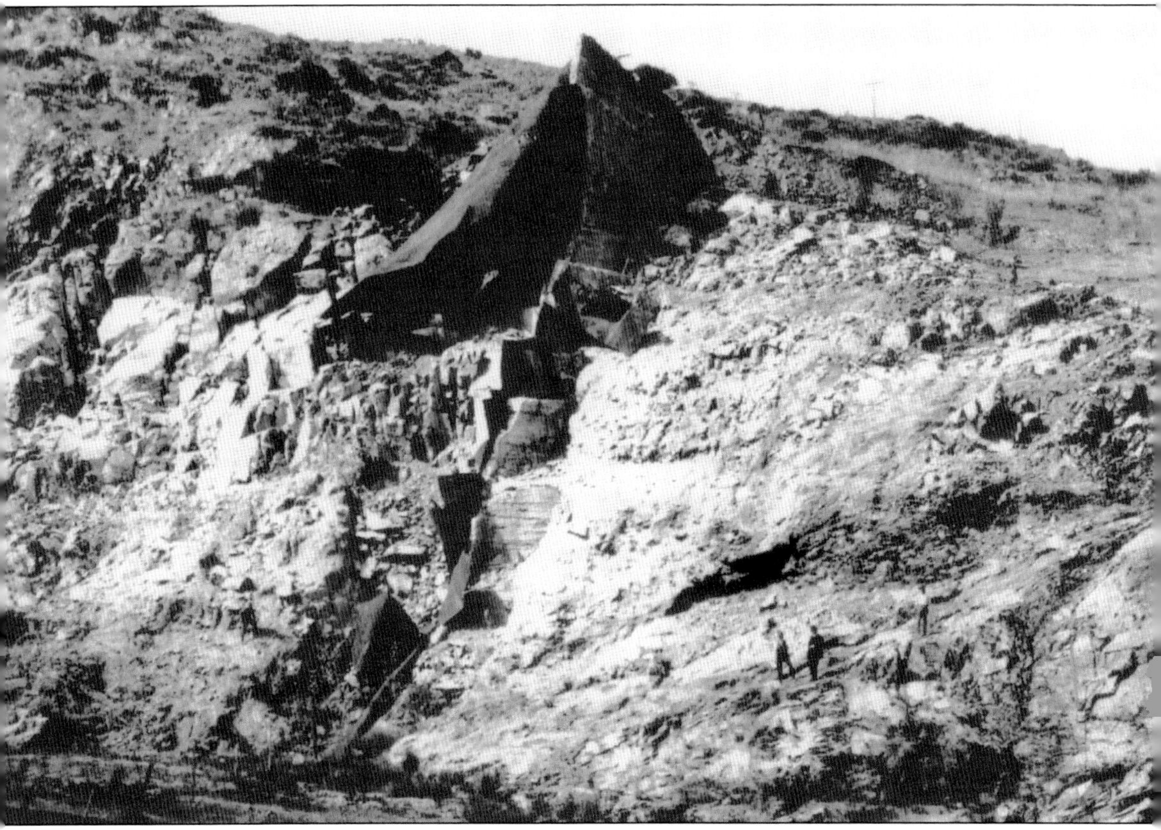

After standing for over 20 years, the rock-filled Lower Otay Dam failed in January 1916. At the time, most people blamed the rainmaker Charles Hatfield for the disaster, but the problem was the dam itself. It simply was not constructed to withstand an episodic flood event. Most experts pointed to the inadequate spillway, which at the time of construction many considered excessive. That winter, a series of storms stacked one on the other marched through the Southland and dropped a prodigious amount of rain over a two-week period. Otay Dam was overtopped, an abutment on the Sweetwater Dam gave way, and the Tijuana and San Diego Rivers flooded. The 20-foot wave that swept through the Otay Valley destroyed ranches and farms and killed 14 people. This photograph, taken immediately after the dam's failure, shows the remains of the west abutment and steel plates projecting from the canyon wall. The flood swept some of the steel plates all the way to the coast. (SDCL.)

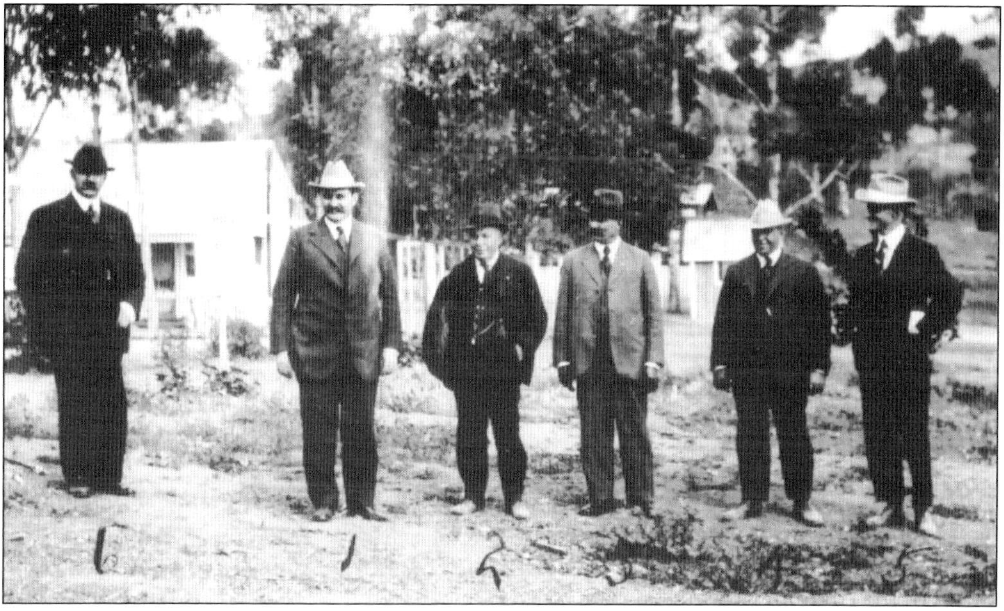

The loss of the city's principal water source compelled city leaders to take immediate action to rebuild the dam. But in an odd circumstance, in November 1916, city voters rejected a bond issue of $628,000 for the reconstruction. It took a second ballot in February 1917 to approve and fund the project. The San Diego Common Council awarded the contract to rebuild Lower Otay to contractor John P. Kennedy. Savage joined Kennedy, William Bailey, C.J. Schmidt, W.J. Hahn, and J.C. Beer at the dam site to sign the agreement on October 11, 1917. Savage soon became dissatisfied with the workmanship and the dilatory pace and convinced the common council to cancel Kennedy's contract and allow him to take over the project. Savage brought in his staff and a crew of day laborers and set to work. (Both, SDCL.)

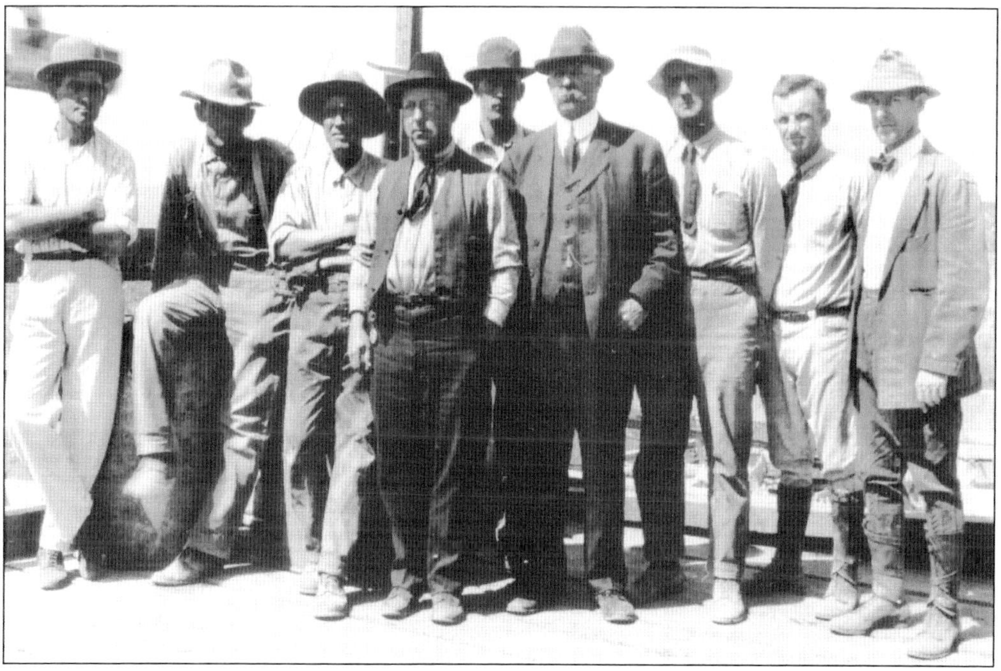

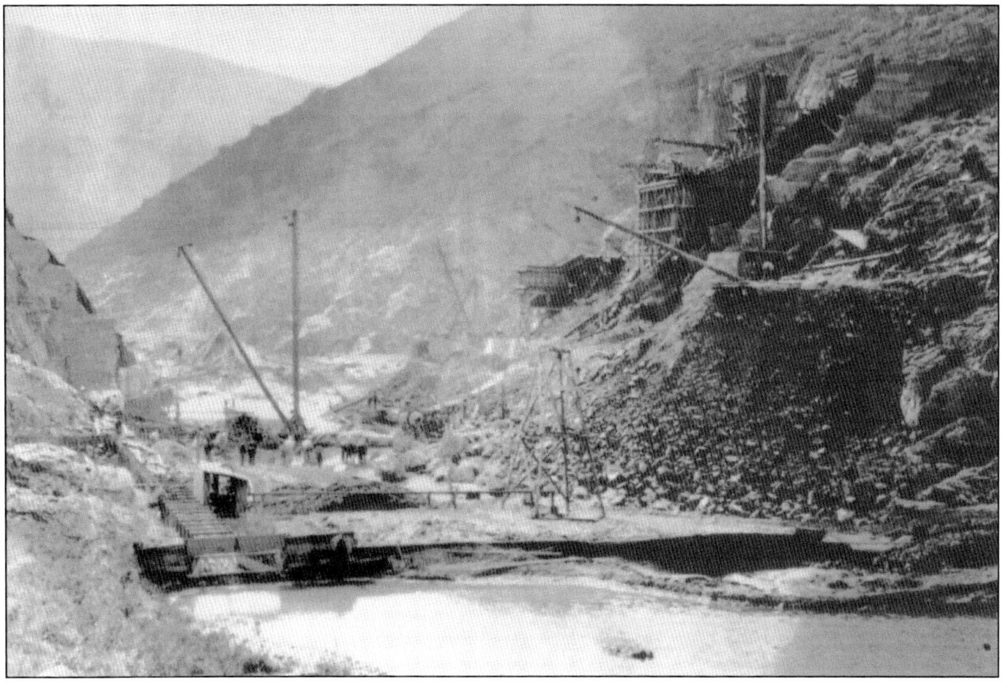

Lower Otay was so distant from town that Savage's first task in 1917 was to build an extensive construction camp to house the laborers, his administration staff, and supervisors. Savage was able to use the foundations of the old dam, which were in good condition and still in solid rock, and the old quarry. Savage even placed the new outlet tower over the site of the old one. The new dam was a gravity section of "cyclopean" concrete, also considered a radial curve dam because the upstream face of the dam was shaped with a 325-foot radius. The dam stood 150 feet high with a 33-foot-wide crest that ran along its 700-foot length. The dam created a reservoir of 56,300 acre-feet. During construction, workers placed over 84,000 cubic yards of concrete in the structure. (Both, SDCL.)

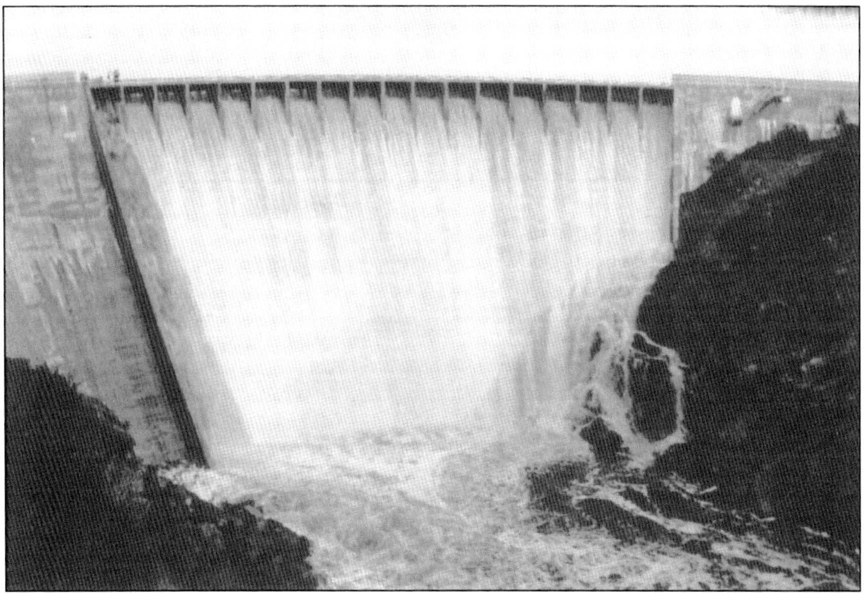

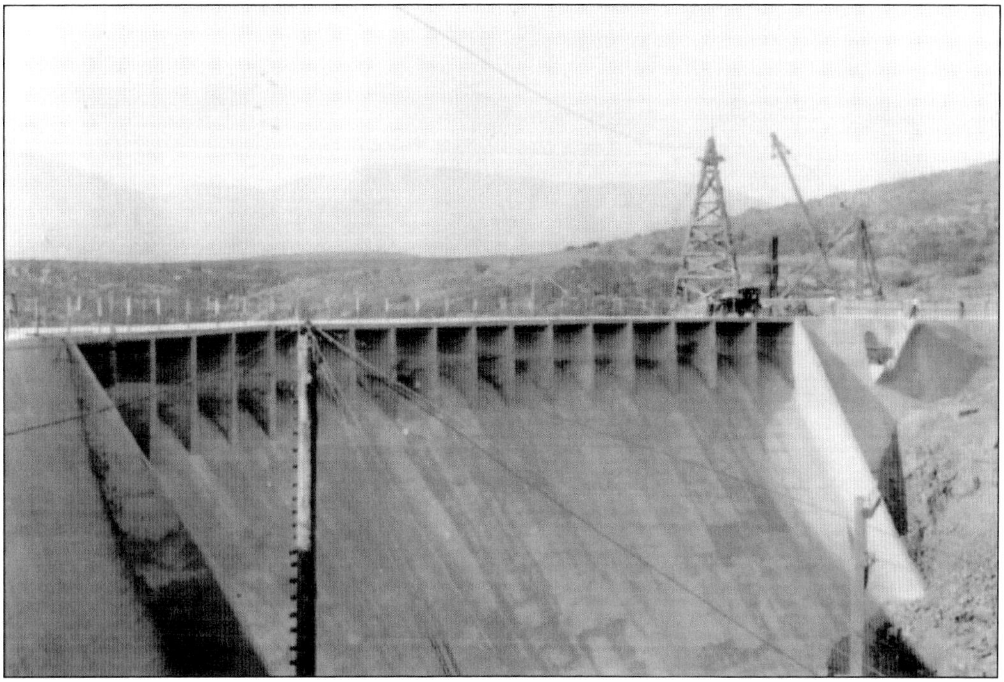

The assumption that the inadequate spillway was the culprit in the 1916 failure prompted special attention to that feature in Otay's reconstruction. The new spillway was a gated overflow system at the center of the dam and an additional open channel on the left abutment. The dam cost $733,157 to build. It is shown above upon its completion in 1918 with the first automobile crossing the crest, allegedly with Savage at the wheel. The city renamed Lower Otay Savage Dam in the engineer's honor in 1934. Savage Dam is seen at right at the exact day and hour 100 years after the flood of January 1916 with pieces of the original steel core plates at the dam's base. (Both, SDCWHA.)

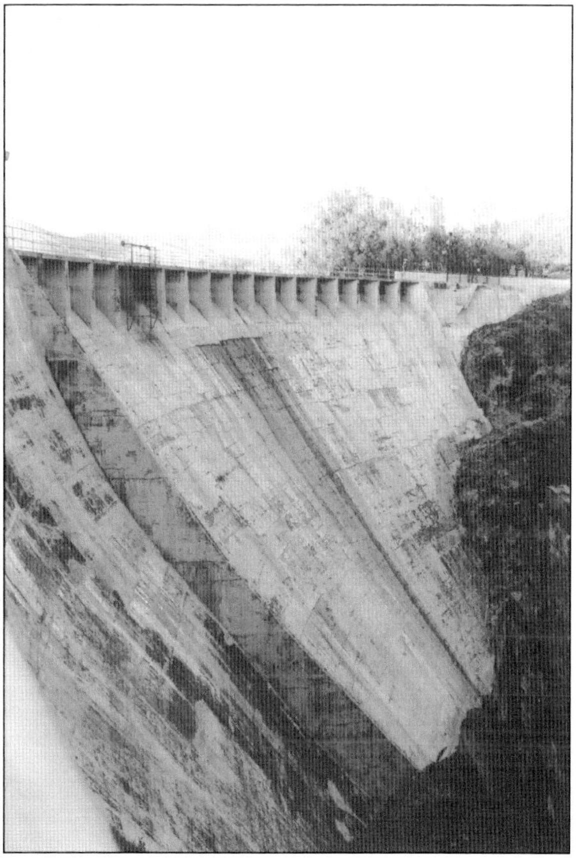

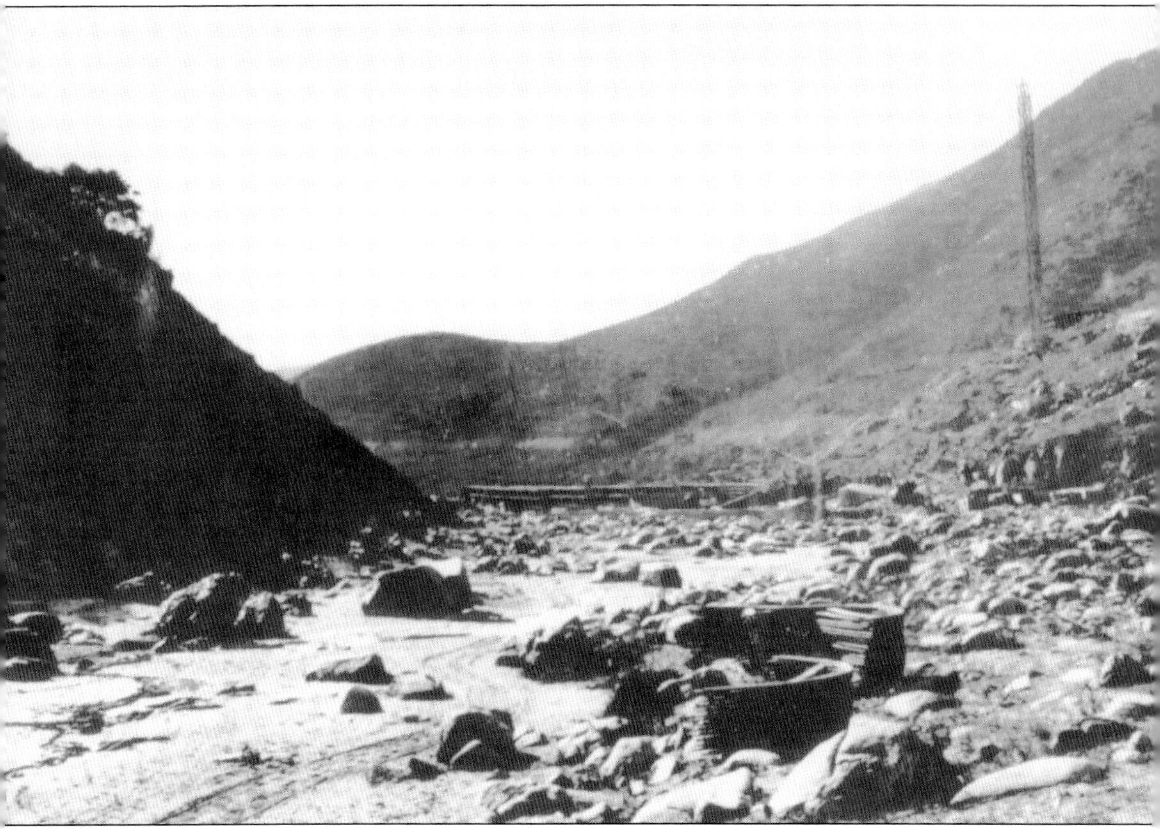

It was trees and not water that initiated the Hodges Dam project. The Santa Fe Railway purchased the San Dieguito Rancho lands to grow eucalyptus trees for railroad ties, but the trees' slow growth and brittle nature ended the plan. At a dinner, Ed Fletcher met E.P. Ripley, the president of the Santa Fe Railway, and Walter E. Hodges, railroad vice president and president of the Santa Fe Land Improvement Company, and convinced them that financing a dam would bolster the commercial expansion of the area. The proposed dam would conserve the waters of the San Dieguito River and be located about five miles southwest of the town of Escondido. The site was in the rocky, narrow, elevated, and sparsely vegetated Carroll Canyon just before the river dropped into the flood plain that passed through Rancho Santa Fe and on to the coast. Being railroad men, not water men, Ripley and Hodges saw the plan as a business venture and agreed on the stipulation that they sell the dam at the earliest opportunity. (SDCL.)

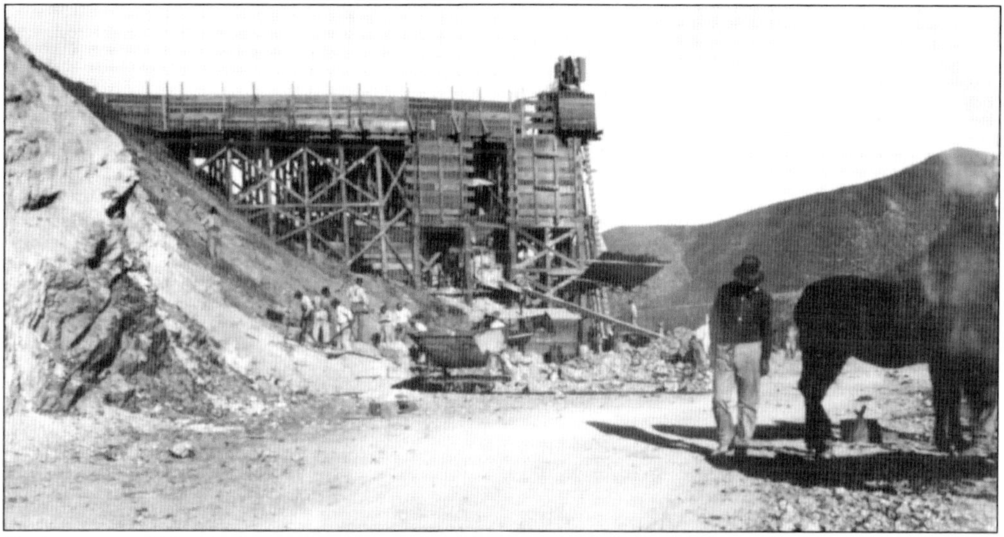

Hodges named Fletcher the manager of the dam project and also of the new San Dieguito Mutual Water Company. Fletcher immediately negotiated the purchase of the Carroll Canyon property. Throughout construction, everyone referred to the structure as the Carroll Dam. The rumor was that if Hodges built the dam, in return for all the favors, Fletcher would see that it was named after him. (SDCWHA.)

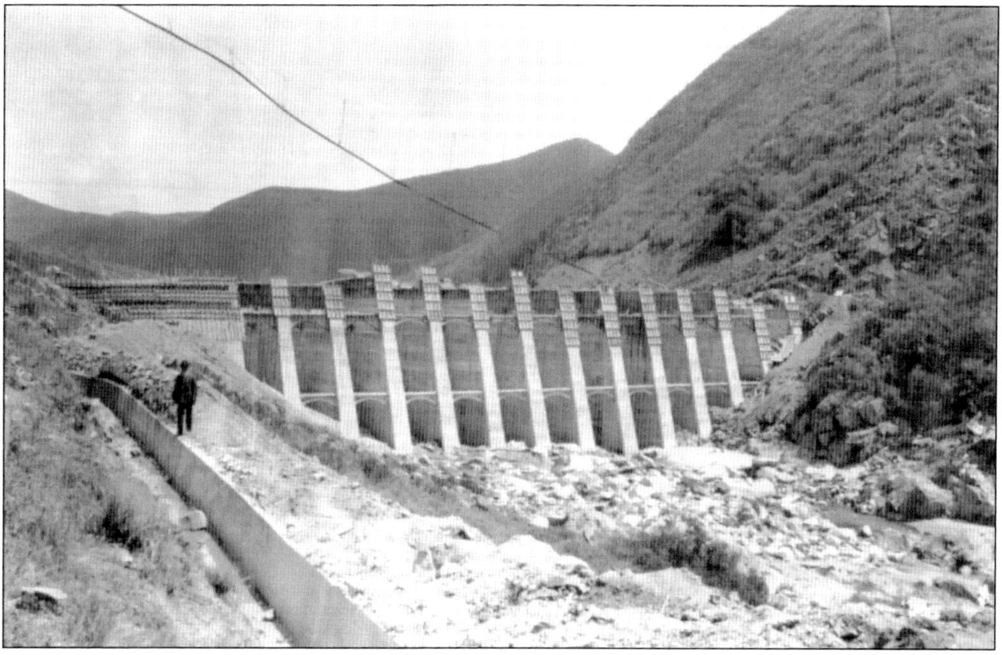

Fletcher contacted innovative dam builder John Eastwood to design a multiple-arch dam for the Carroll site. Ironically, Eastwood had plans drawn for a similar dam for the Volcan Water Company. In 1918, Fletcher, representing the San Dieguito Mutual Water Company and the Santa Fe Land Improvement Company, selected the Bent Brothers Construction Company, which had worked on the Sweetwater Dam improvements in 1911, to undertake the project. (SDCL.)

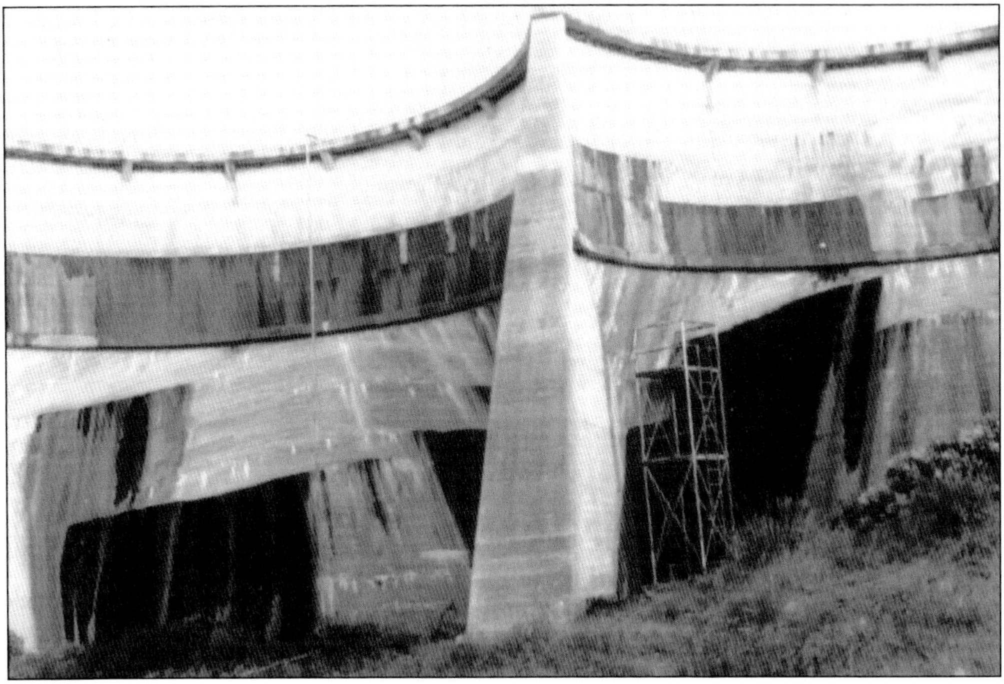

Bent Brothers started pouring concrete in November 1917 and finished the project 12 months later. According to Eastwood's plans, workers reinforced the concrete arches with steel rebar. As the arches grew taller, they were inclined from 45 degrees in the last 10 feet to vertical at the top. The engineers strengthened the dam with concrete buttresses braced with horizontal struts, and installed a solid toe along the downstream side. Interestingly, and concerningly for some engineers, Eastwood did not order the buttresses reinforced with steel. Shrinkage and temperature stresses caused cracks in the unreinforced buttresses shortly after the dam was completed. A report from the state inspection board said the buttresses were sufficient but thought the spillway was inadequate. (Both, SDCL.)

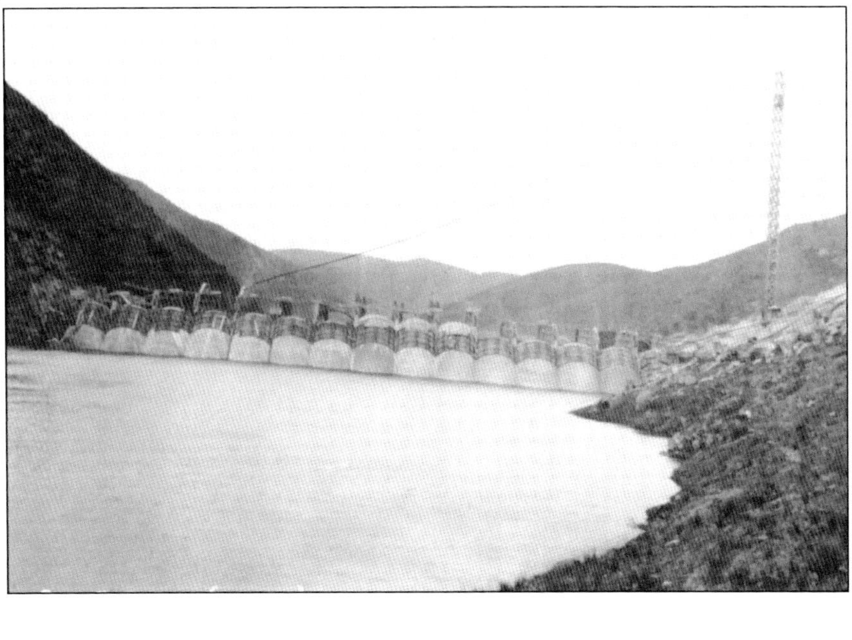

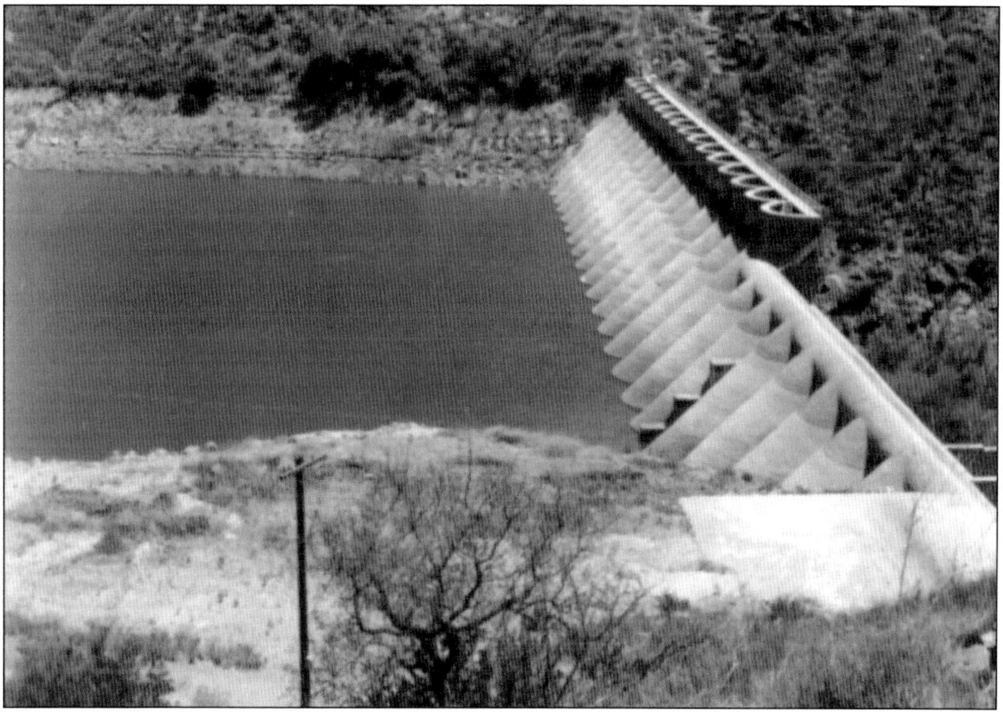

Hodges was a gravity multiple-arch dam consisting of 23 hollow, 24-foot-wide, 24-inch-thick reinforced concrete arches supported with buttresses of mass concrete. The 550-foot-long, 137-foot-high dam cost $328,463. Because the project was private and not a public utility, state engineer Wilbur F. McClure intervened before the completion of the dam and ordered a new spillway engineered. The redesign placed a new, larger spillway on the northern abutment, which extended the dam an additional 360 feet and reduced the crest almost 15 feet, which in turn decreased the overall reservoir capacity. The spillway improvement proved prescient, as successive years of excessive rainfall produced notable flood events in the county. Below, from left to right, Hodges, E.P. Ripley, and Fletcher and their wives visit the dam just before completion in December 1918. (Above, SDCL; below, UCSD Special Collections.)

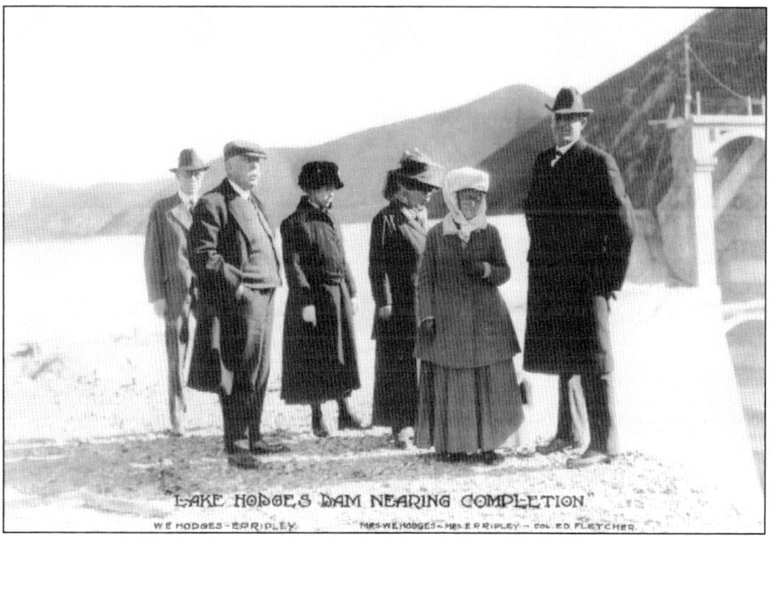

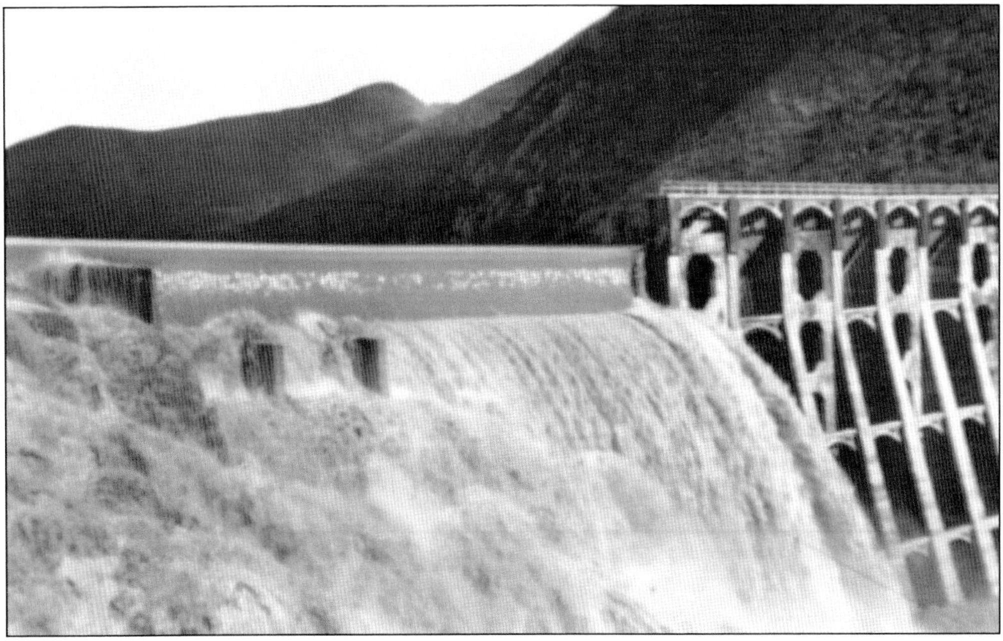

The worthiness of the multiple-arch concept concerned some hydraulic engineers, but to Eastwood's credit, even an incomplete Hodges withstood the strong rainstorms and flooding that overtopped the dam in 1918. *Western Engineer* magazine remarked on the structure's stability and reported that the arches were watertight even without the final concrete coating. The analysis proved valid, as the dam easily weathered the severe storms that struck the north county in 1922 and 1927. In 1949, the California Division of Water Resources recommended that a larger dam, referred to as Super Hodges, be placed downstream, which would have submerged Hodges, but the project received little support and never materialized. The photograph below shows, from left to right, Ripley, Hodges, and Fletcher at the dam in 1918. (Above, SDCWHA; below, UCSD Special Collections.)

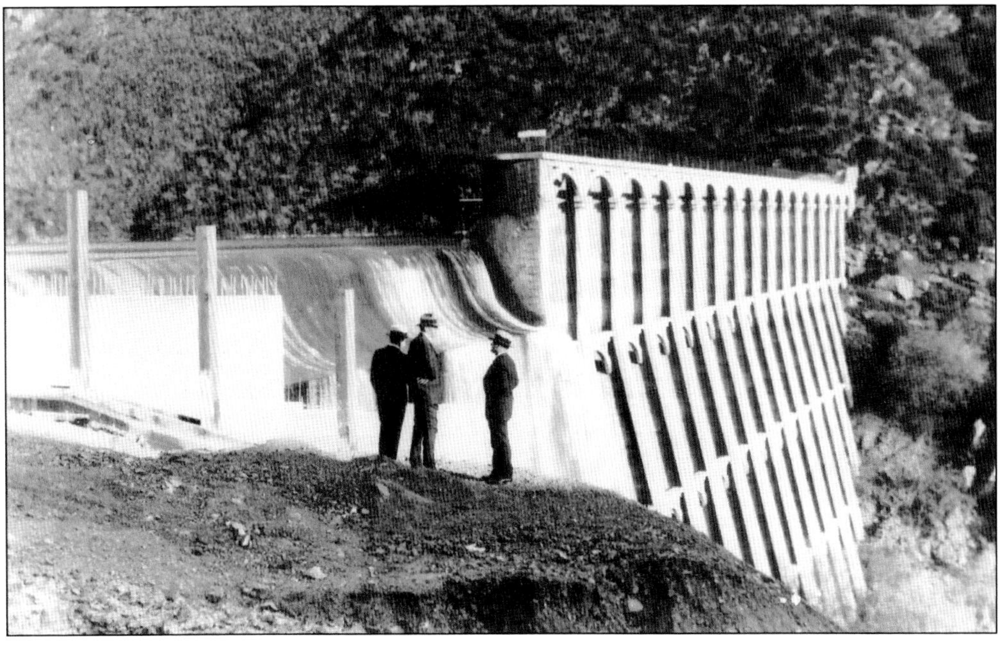

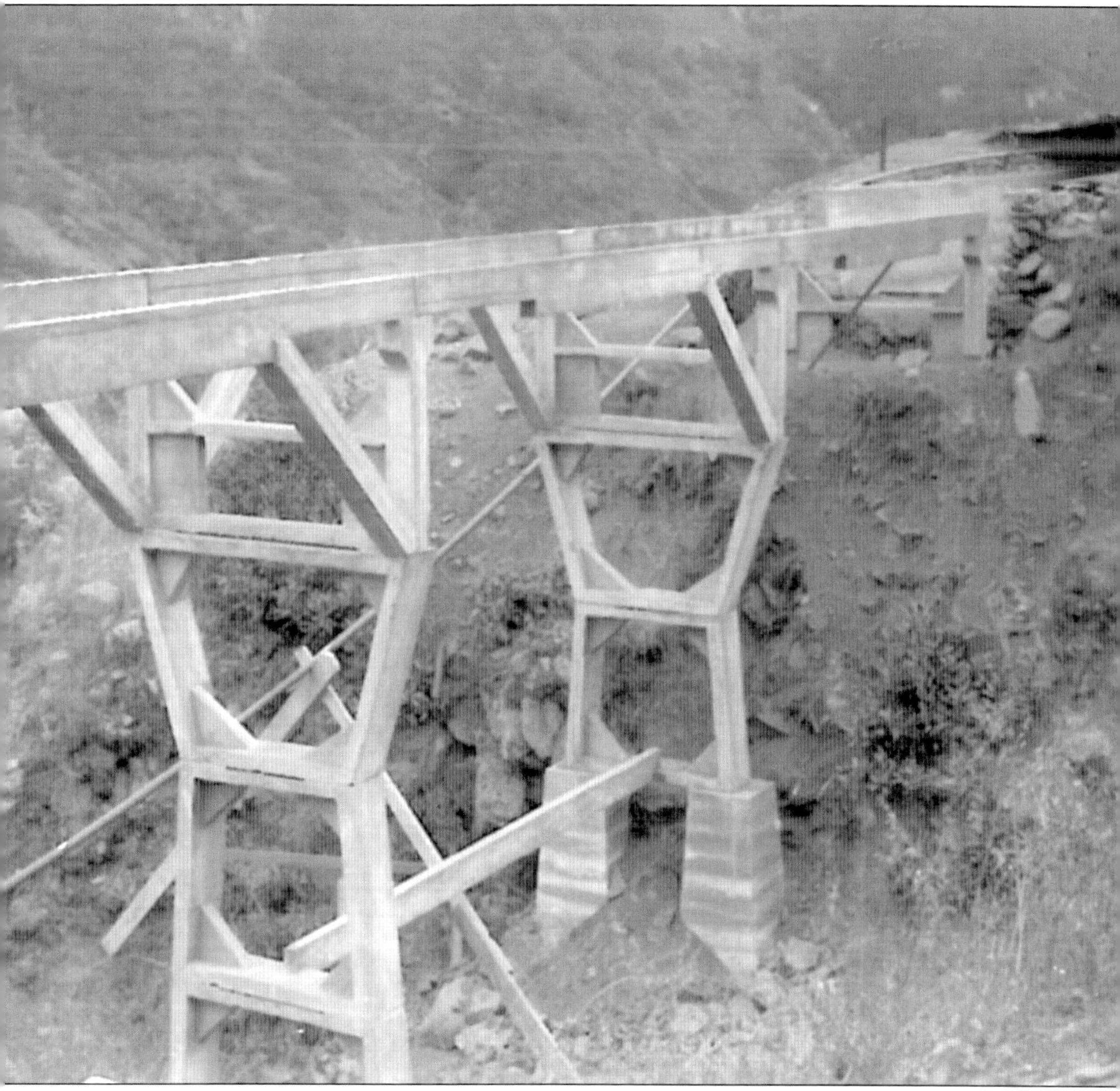

An element of the Hodges project included transporting water from the Hodges Reservoir to the new San Dieguito Reservoir. As Bent Brothers constructed Hodges, the firm was also building San Dieguito Dam several miles to the west. The contractors designed and constructed a four-mile concrete-lined, open gravity-flow flume to connect the reservoirs. The water in the conduit passed through a four-foot-deep concrete ditch that crossed the flat below the dam, into several tunnels, then into an open metal-framed channel sitting on concrete trestles that crossed the small gullies. A series of siphons, closed pipes with the upstream end at a higher elevation than the downstream end, pulled the water through the narrow intervening canyons. The flume was originally called the Carroll Flume, but in time it became the Hodges Flume or the San Dieguito Flume. It remained in use until Santa Fe Irrigation replaced it with a pipeline in 2004. (SDCWHA.)

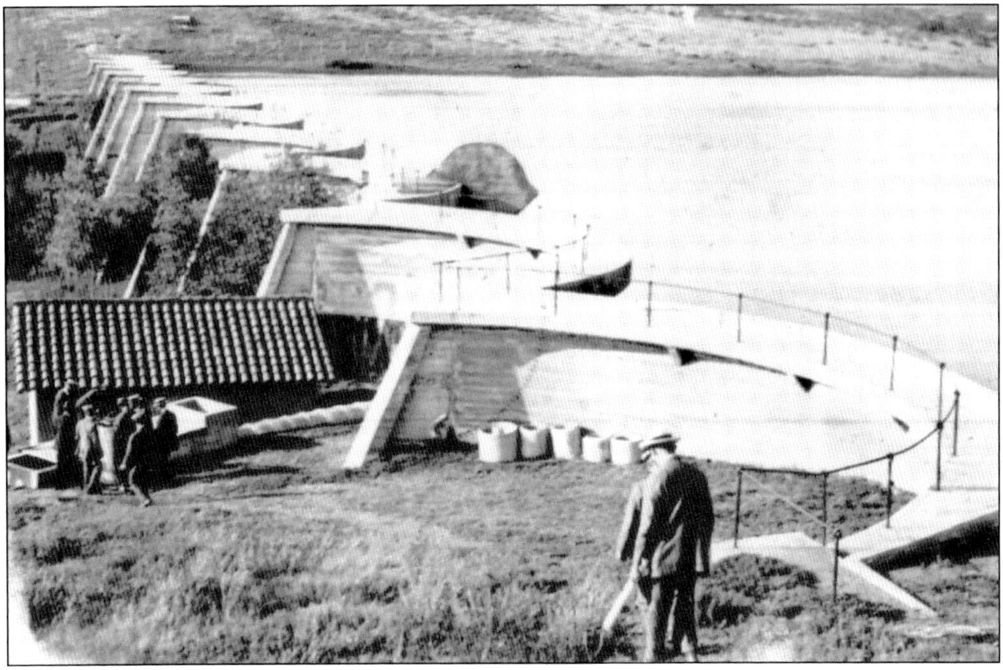

The Santa Fe Railway consortium authorized Bent Brothers to construct San Dieguito Dam to act as a regulating reservoir and distribution facility. Constructed in only four months in 1918, the 51-foot-high concrete, hollow, gravity, multiple-arch structure had a four-foot crest running its 650-foot length and cost approximately $160,000 to construct. The dam's 11 arches are spaced at 50-foot intervals between the centers of the buttresses. Engineers capped the crest with coping, which they raised one foot on the upstream side to create both a wave parapet and a walkway. The spillway was a siphon in the third bay from the left abutment with its top at the level of the crest. The view above from the south end of the dam was taken during a 1918 inspection tour. (Both, SDCWHA.)

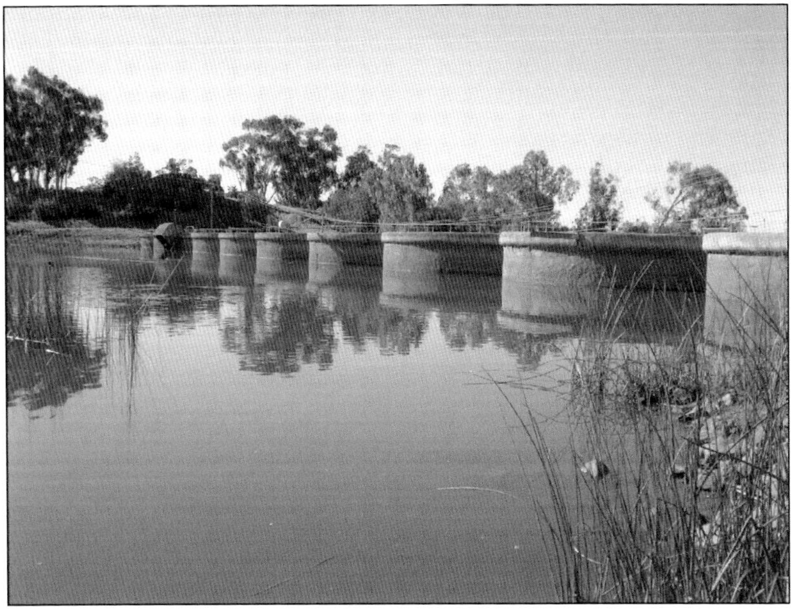

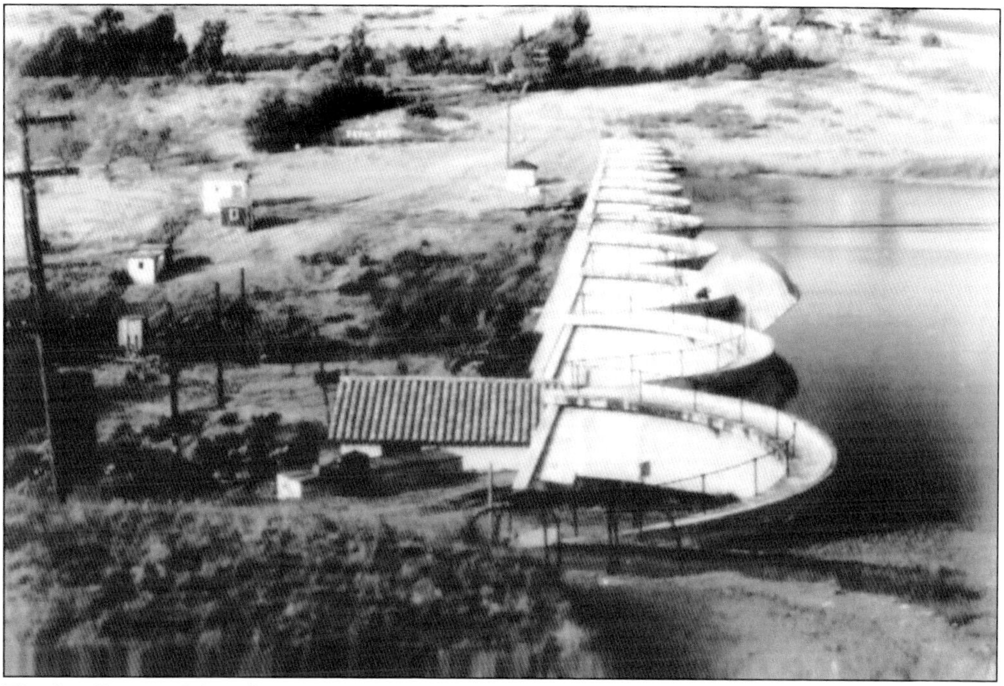

The San Dieguito Mutual Water Company constructed the original small dam to conserve Escondido Creek in the Old San Dieguito Rancho in 1915. The dam was amid the citrus and avocado orchards of the community of Rancho Santa Fe, about eight miles from the coast. The water company owners also hoped the small reservoir could supply water to the growing San Dieguito towns of Encinitas, Cardiff, and Solana Beach. A redwood-stave pipeline served as the distribution system to feed the water to the coastal towns. In most years, the tiny catchment left the small reservoir as more of a picturesque lake than the core of a water system. The small earth dam remained in place until Bent Brothers built San Dieguito Dam in 1918. (Both, SDCWHA.)

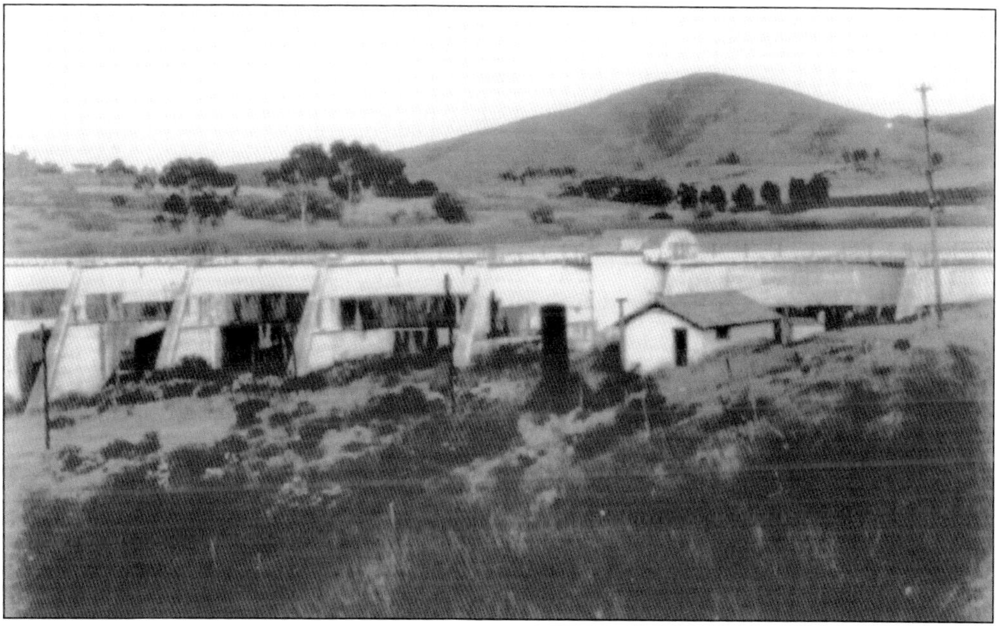

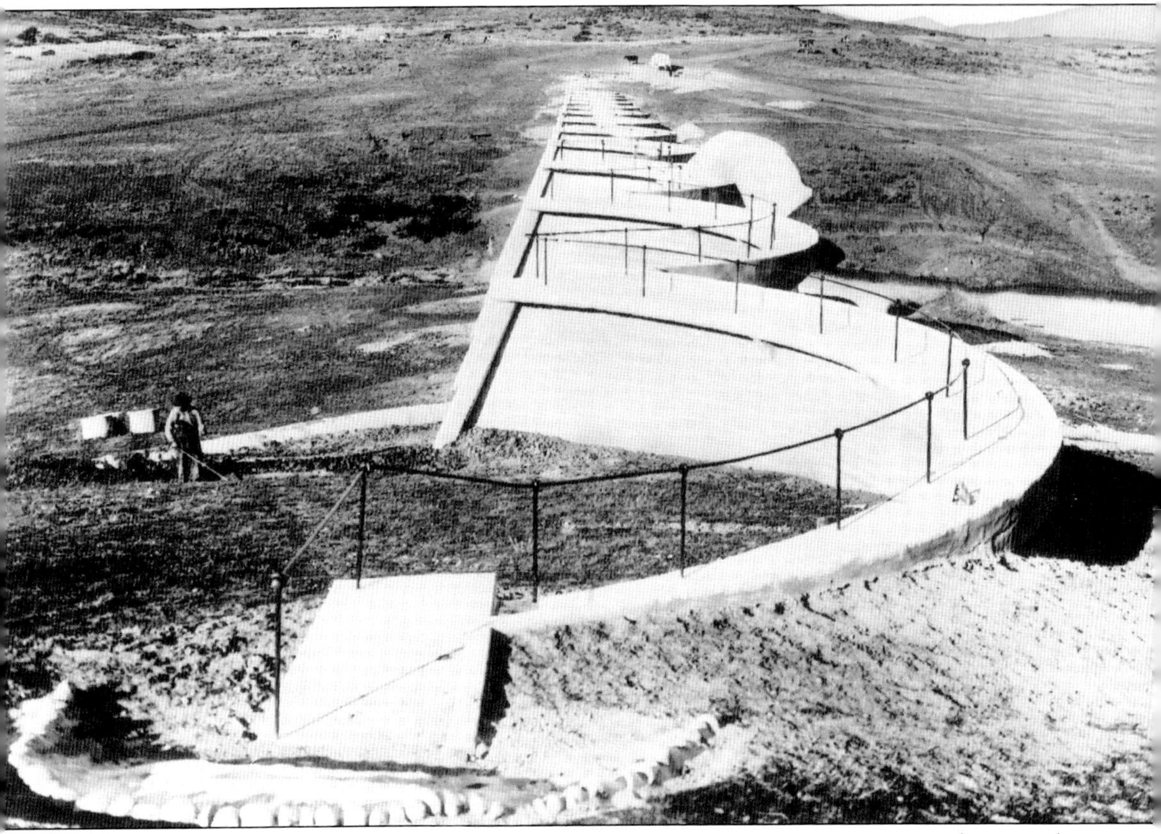

When Fletcher selected the Bent Brothers Construction Company, its most recent job was working on the reconstruction of the concrete gravity Sweetwater Dam in 1917. The engineers understood that arch dams depended on the horizontal shape of the arch to transfer the pressure the water in the reservoir exerted downward into the dam's foundation, so to strengthen the arches and stiffen the overall structure, they ordered Bent Brothers engineers to place auxiliary buttresses on each side of the main buttress rather than the usual method of mounting horizontal struts between them. The dam's relatively small height made this possible. San Dieguito Dam has stood largely as constructed in 1918 until early 2017, when Santa Fe Irrigation opened a project to replace the original, aged pump station. (Courtesy of UCSD Special Collections.)

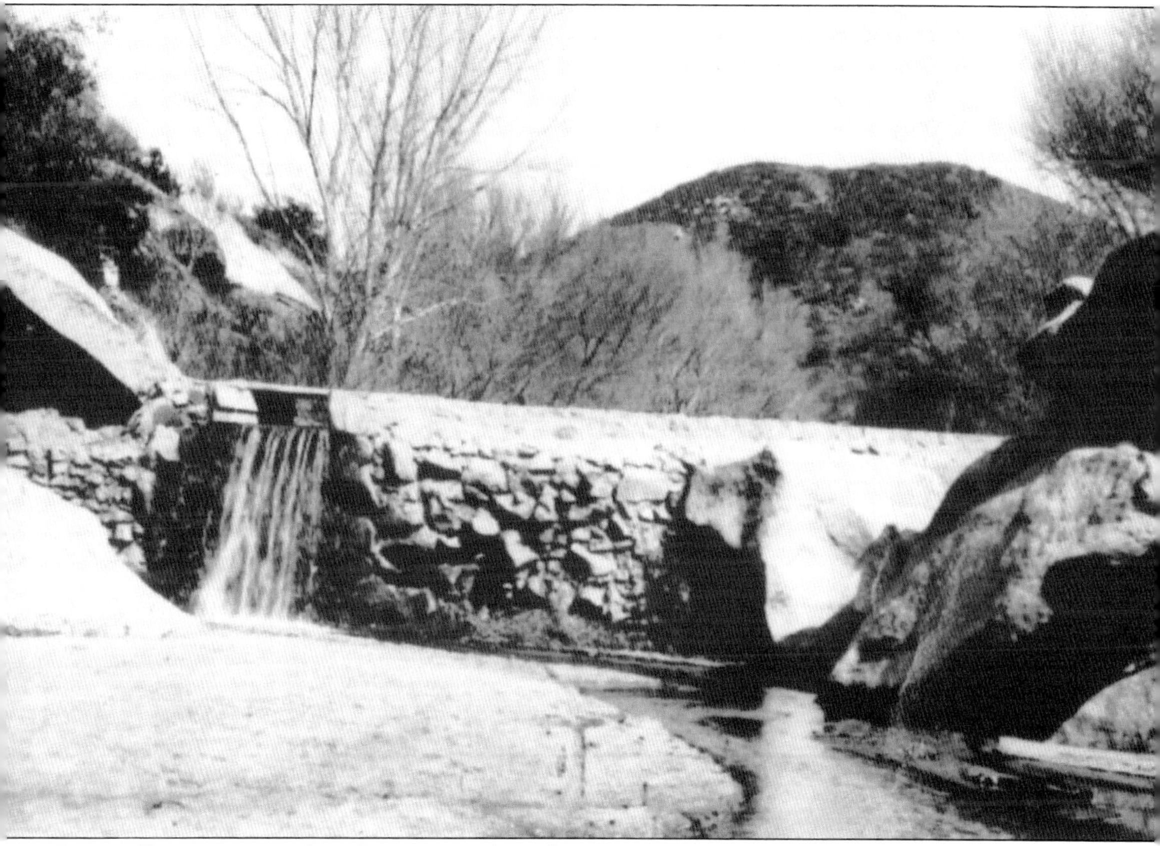

Barrett Dam originated in the 1890s, when the SCMWC designed the small masonry and concrete diversion dam shown here in 1897. The site was in a narrow canyon at the confluence of Cottonwood Creek and Pine Valley Creek. The company owner, John D. Spreckels, hoped to feed water from the dam via a conduit to the Lower Otay Reservoir. In the end, he abandoned the project because the dam was too small to create a reservoir with sufficient supply to survive a drought, and he discovered several significant structural issues with the dam's foundation. At the time Spreckels recognized the project's limitations, he was already in negotiations with the city over the sale of Lower Otay, so he gladly relinquished the dam site and his entire water system to the city in 1912. As it happened, San Diego city engineers had already identified a nearby site that could sustain a larger structure. (Courtesy of Vista Irrigation District.)

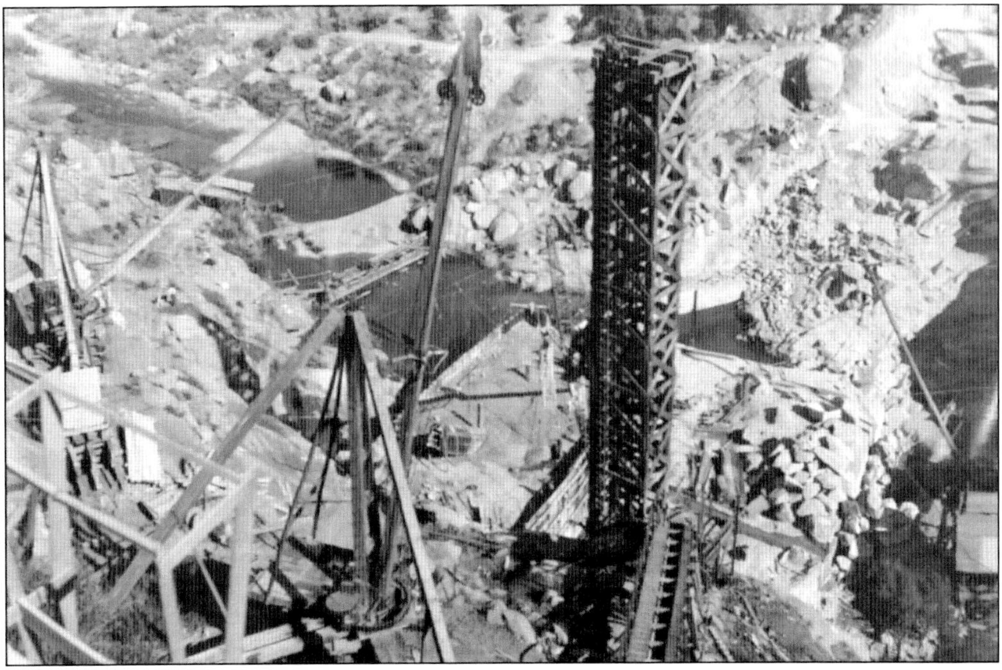

City engineers had identified a new Barrett site, but they had no defined plans for developing it until the 1916 flood event overtopped Lower Otay and overflowed Morena. Now city water experts recognized the need for better flood control in the river valley and saw an opportunity to conserve future floodwaters. In 1919, city hydraulic engineer Hiram Savage opened the project with W.A. Stebbins as the construction engineer. Because the dam was in a remote locale almost 35 miles south and east from town, Savage was compelled to create an extensive construction camp to support the workers and his staff, many of whom lived on site for nearly two years. This 1920 view of the camp and aggregate pits was taken from the quarry road. (Both, SDCL.)

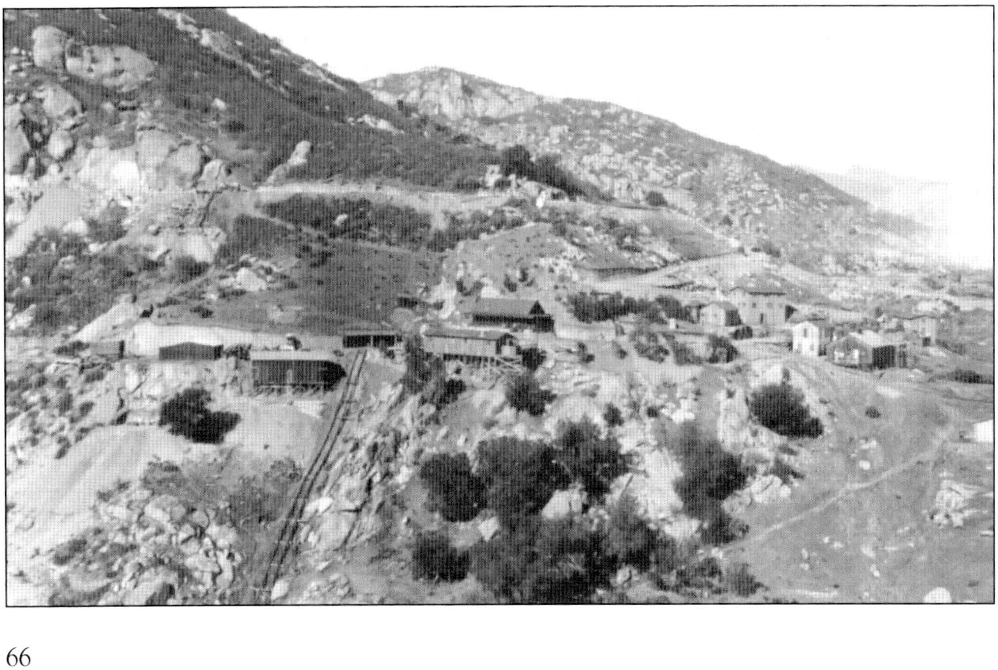

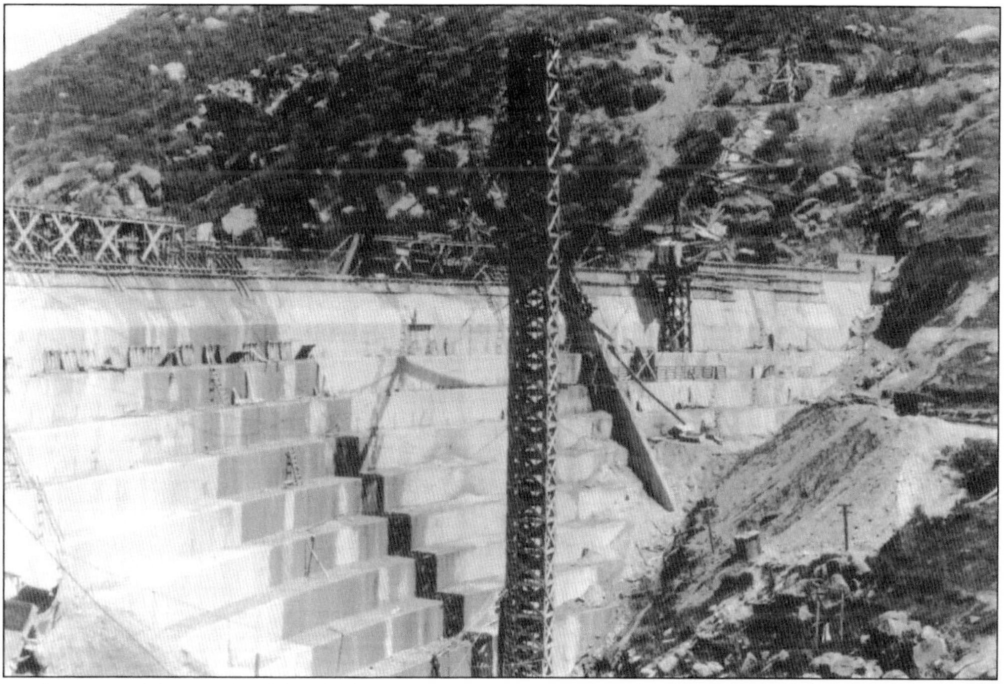

Unforeseen delays pushed back Barrett's starting time into December, when rains arrived and floodwaters threatened the project. Rising waters behind the construction site submerged forms and forced workers to place sluiceways and concrete-supported wooden flashboards to divert the flow and prevent the waters from reaching the construction zone. For months, laborers toiled on three shifts, with the work taking place as much as 10 feet below the level of the diverted water. The heavy rains made the roads impassable and forced delivery trucks to detour 25 miles to reach the job site. Adding to the woes, engineers discovered they would need to perform an extreme excavation on the left abutment where the granite bedrock had decomposed. *Engineering News* aptly described the circumstances in an article titled "Completion of Barrett Wins Close Race with Water." (Both, SDCWHA.)

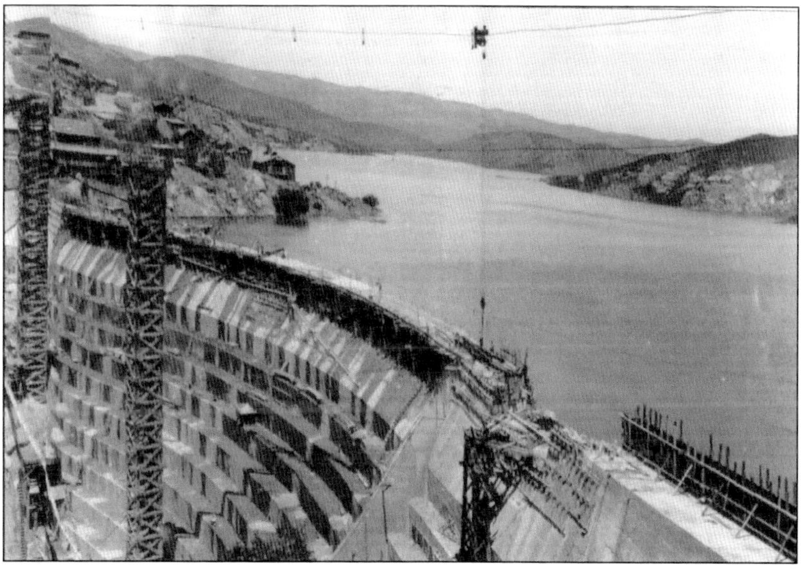

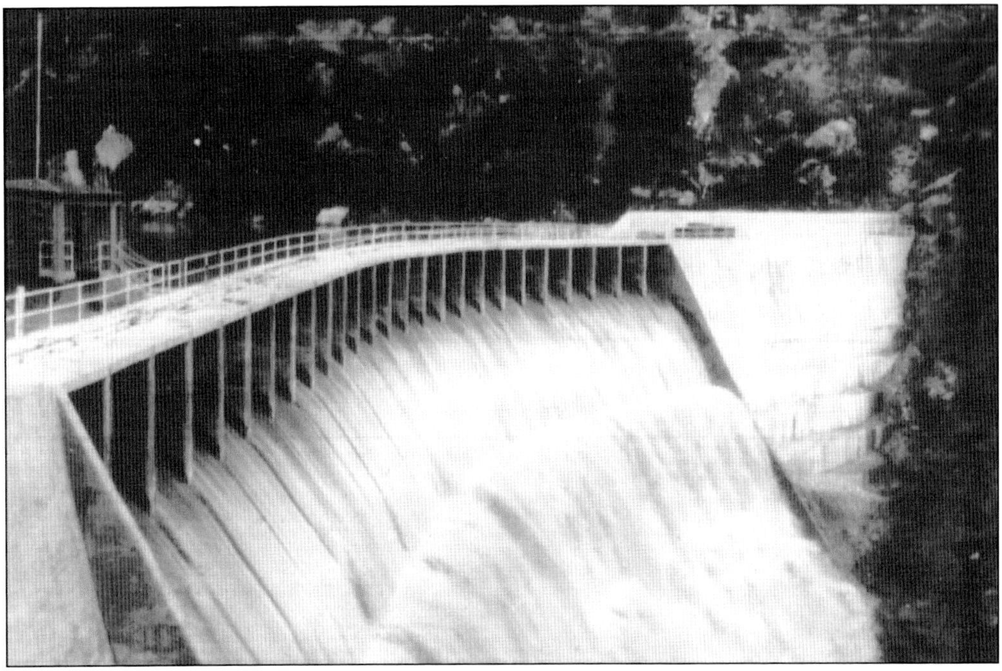

The city dedicated Barrett Dam in July 1922. The dam cost $1,568,353, an amount an engineering periodical said the city would recoup in water savings within a few years. The city named the structure after George W. Barrett, who settled in the area in 1879. The concrete arch gravity section dam stood 171 feet high, with a 15-foot crest running along the 746-foot length. (SDCWHA.)

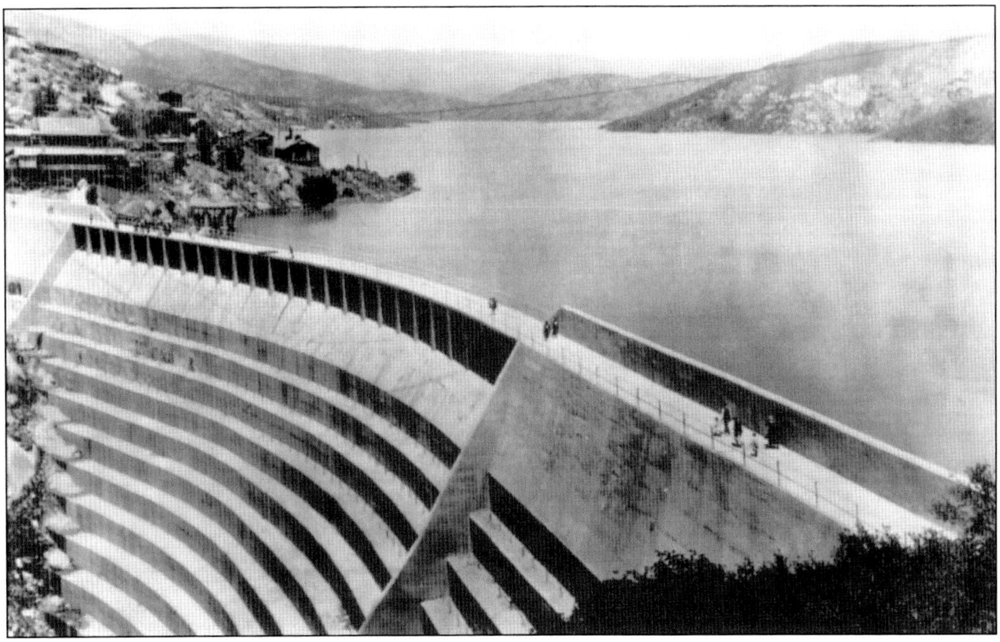

City engineers designed the unique stepping feature across the downstream face to allow for future heightening. The dam had a capacity of 34,200 acre-feet, an 891-acre surface area, and gathered water from a 252-square-mile catchment. City engineers estimated the reservoir would deliver 4.8 million gallons of water to the city every day. (SDCWHA.)

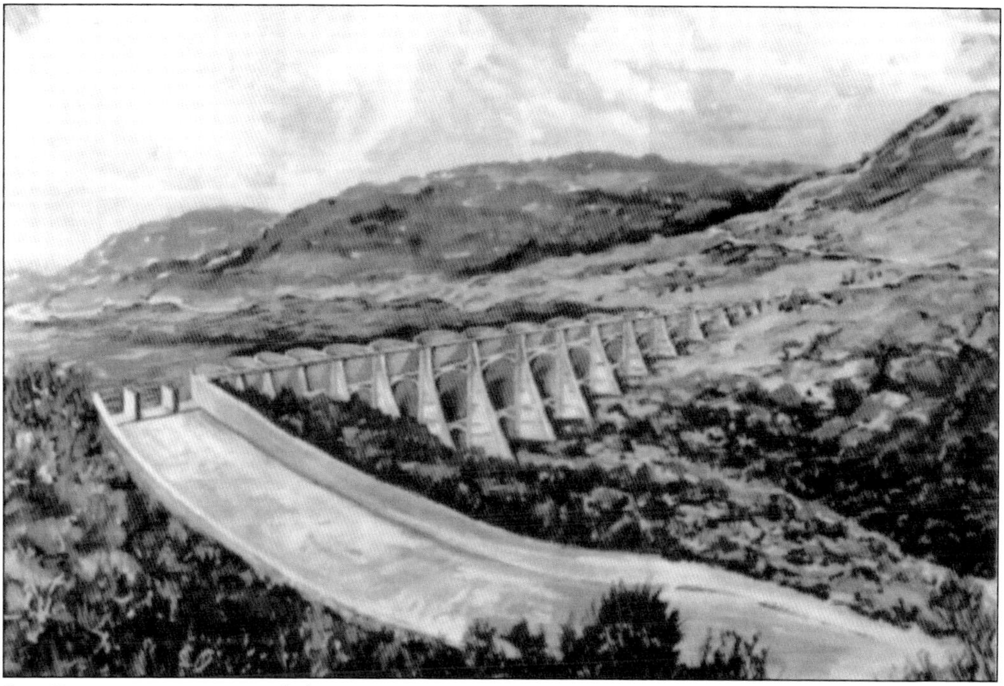

San Diego voters approved the construction of Sutherland Dam, a multiple-arch dam on the property of John Sutherland, in October 1926. This conception drawing shows the first plan for a site 10 miles northwest of Ramona where a dam could collect the runoff waters from Santa Ysabel Creek and the San Dieguito catchment. With runoff meager in those catchments, the dam would realistically serve as a diversion unit for delivering its stored waters to the projected San Vincente Reservoir. Unfortunately, work started in the midst of one of the county's wettest winters on record, and the contractors were forced to stop work. Supervising engineer J.W. Williams reopened work later in 1927 but discovered serious issues with the foundation and again halted work. Eighteen months after the $885,000 project opened, it ended. (Both, SDCWHA.)

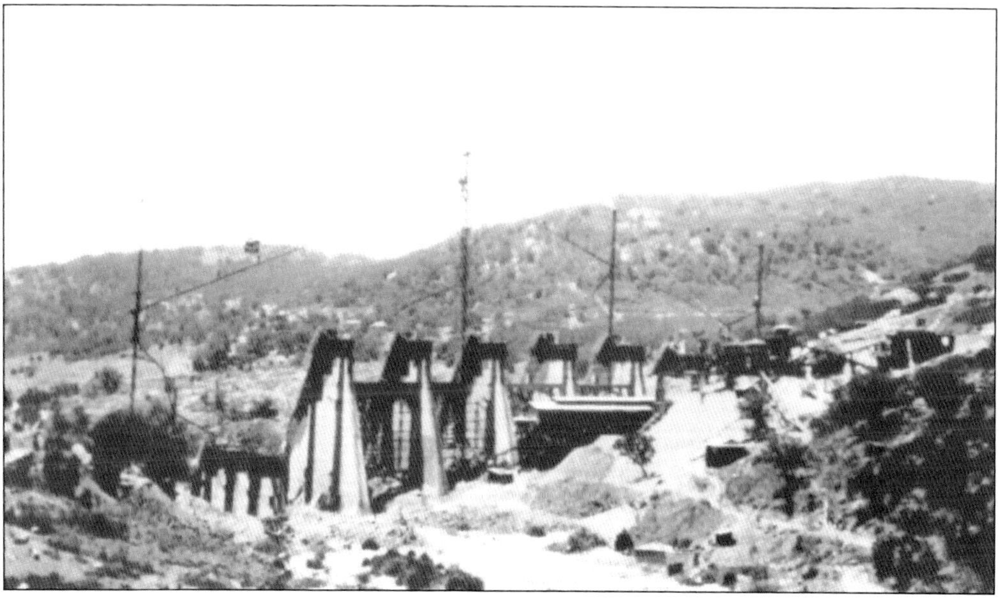

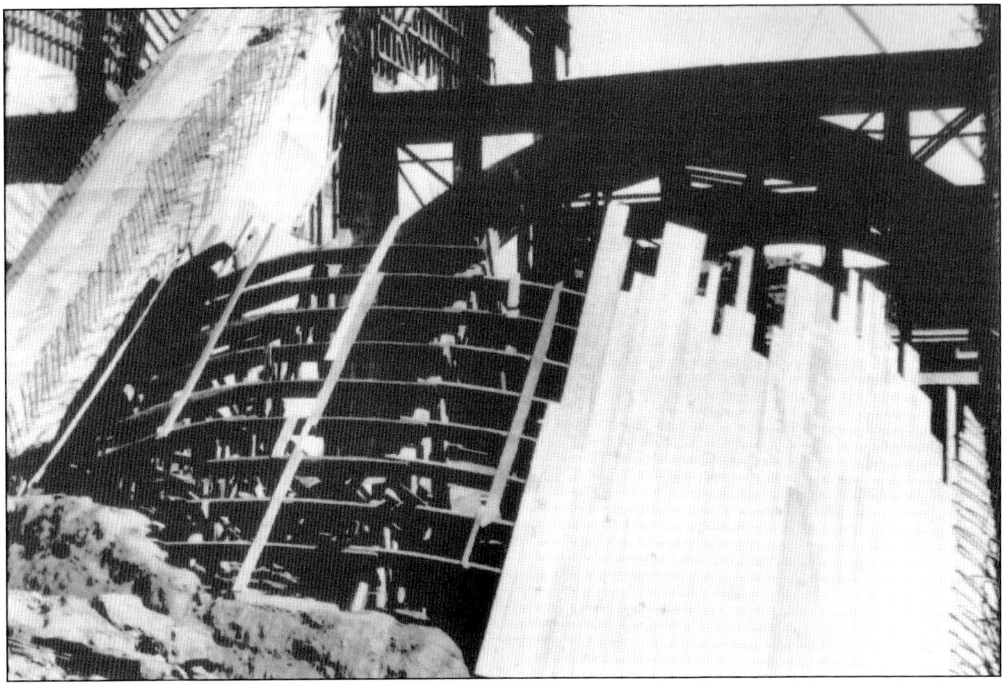

Construction on Sutherland stopped with seven arches poured and most of the wooden frames in place. City engineer C.F. Tolman investigated the work and recommended abandoning the current site and relocating the dam several hundred yards upstream. The new site had a wider span and a stable foundation essential for an arch dam, but the site change also increased the construction time and added expense to the project. (SDCL.)

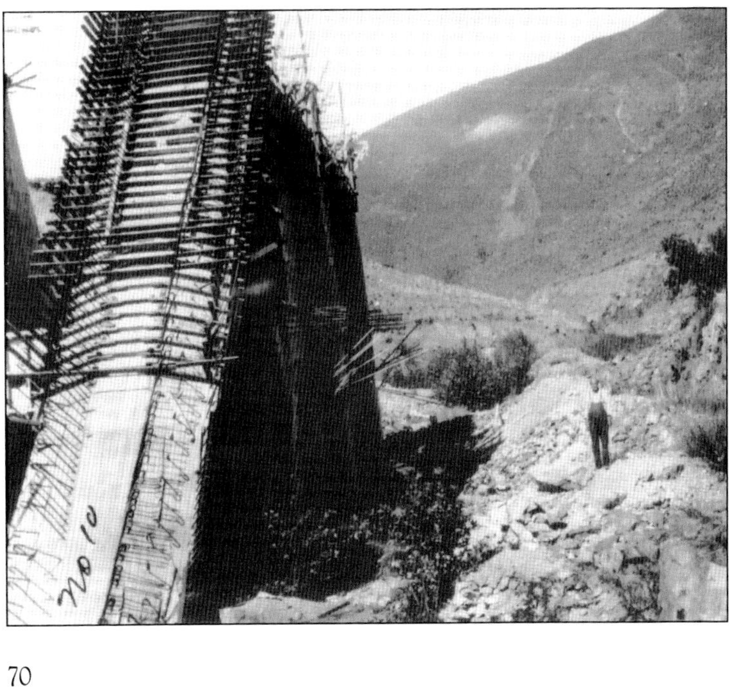

Hiram Savage, fired as city engineer in 1928, encountered this situation when he returned to San Diego after a three-year international hiatus. City attorney Shelly Higgins related that Savage was so opposed to the Sutherland project, and arch dams in particular, that he jumped off the westbound train in Campo, the nearest stop to the dam, rushed to the site, and unilaterally deemed Sutherland unsatisfactory. (SDCL.)

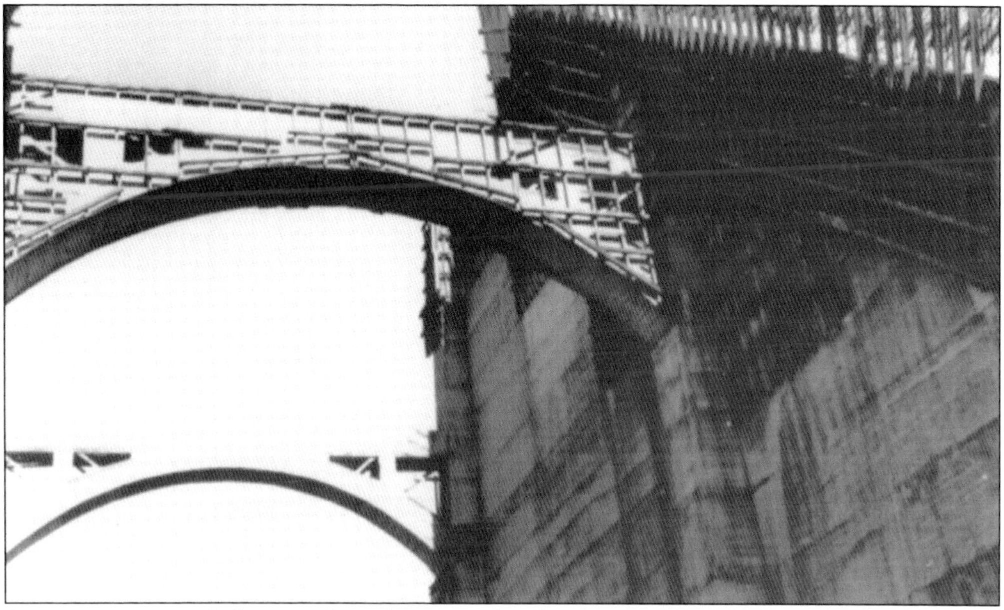

Savage recommended suspending the Sutherland project. The project was a victim of poor engineering, financial problems, and political bickering, but Savage's political foes blamed the project's failure on his dislike of multiple-arched dams. The dam's demise came when the common council transferred all unexpended Sutherland funds to the El Capitan project in 1928. The skeleton of the abandoned dam sat at the site for the next 25 years. (SDCL.)

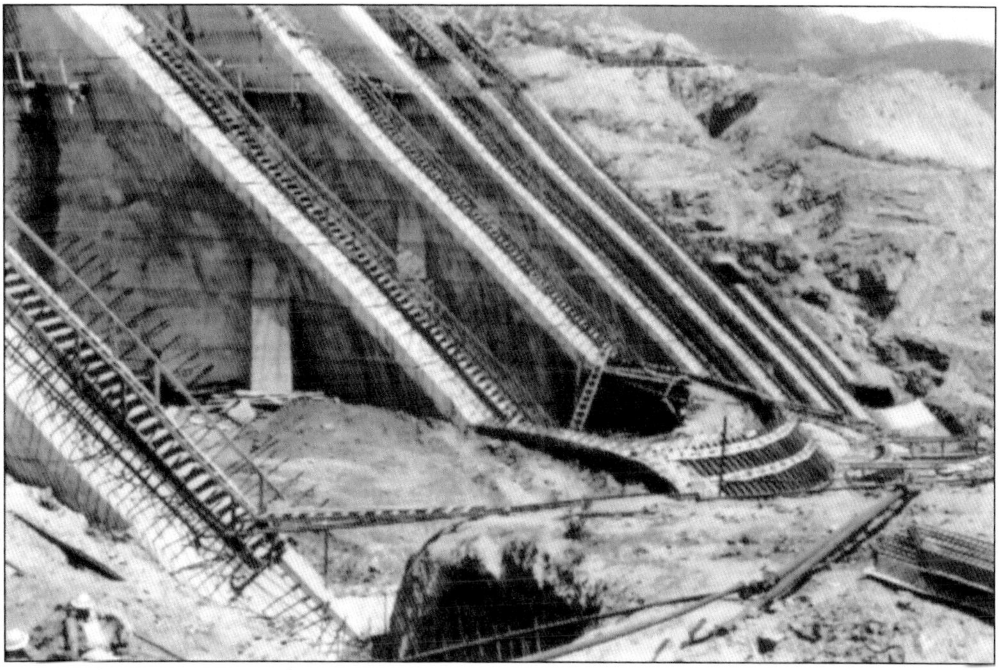

In 1949, the state Division of Water Resources recommended completing Sutherland and using it in conjunction with a new Super Hodges Dam. The report averred that it would be more cost efficient to complete the partially constructed arch dam than to redesign a new earth-fill embankment-type dam. The city voters approved $6.5 million to resume construction in 1952. (SDCL.)

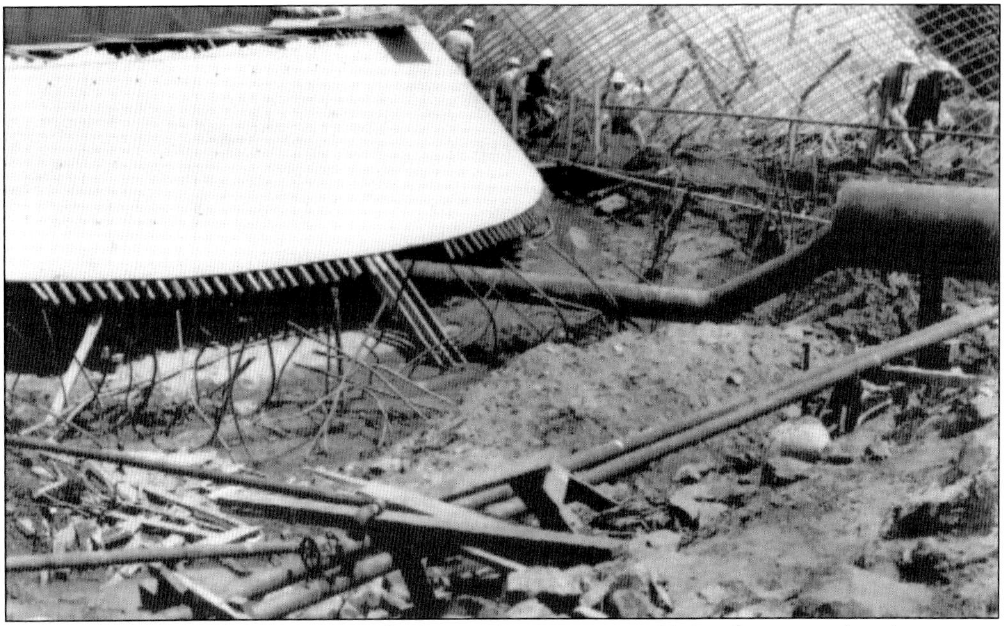

The new Sutherland dam consisted of 17 arches spanning 60 feet between the abutments. The arches were approximately five feet thick at the base tapering to about two feet at the top. Horizontal struts between the 10-foot-thick buttresses provided the strength. Engineers believed the 164-foot-wide, open-channel spillway on the crest would survive any overflow, which has never occurred. (SDCWHA.)

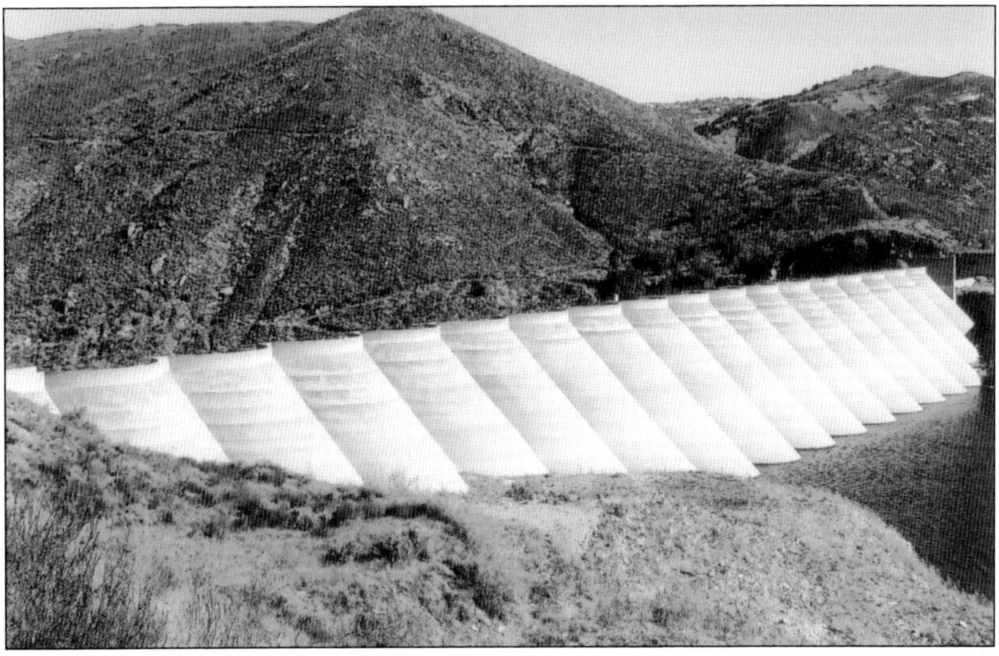

With Sutherland's small catchment, it takes a substantial flood episode to fill the reservoir, which rarely happens. When it does, as in 1980, the reservoir's 556 surface acres conserve 29,220 acre-feet of water, which creates a water supply of 26 million gallons per day. Water in the reservoir passes via pipeline to San Vincente for transfer to the city system. (Author's collection.)

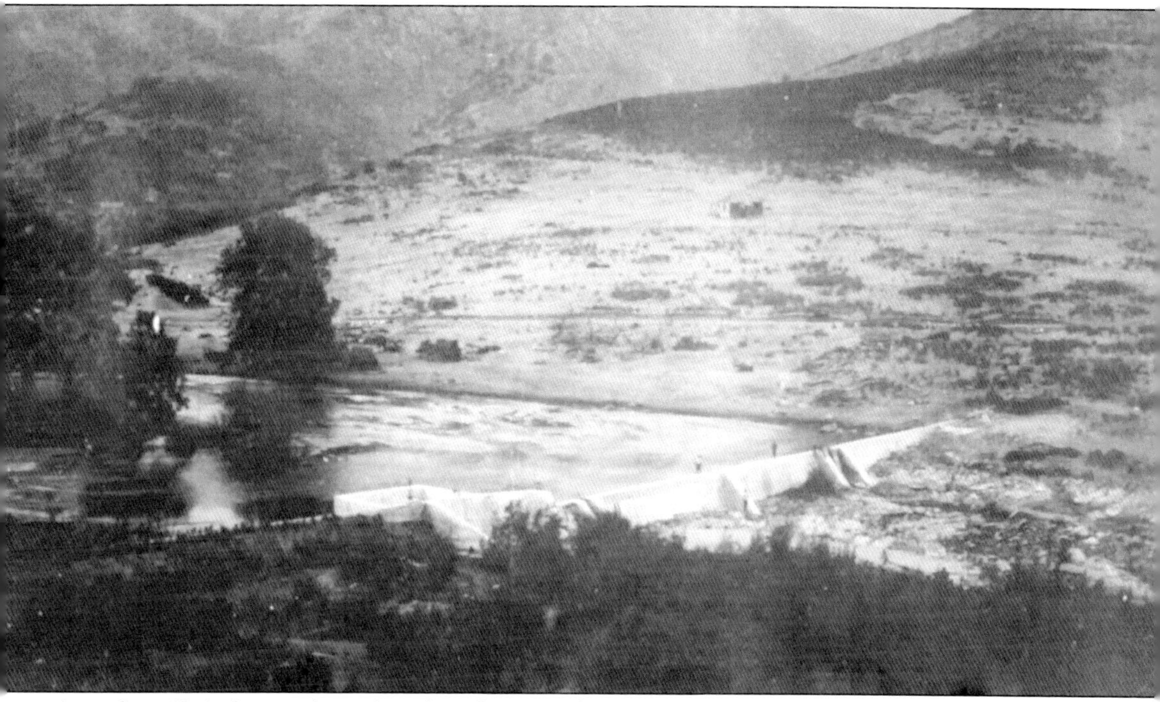

As early as 1914, the city drew plans for a diverting dam to make Upper Otay a more useful facility. In 1925, the city constructed Harvey Diverting Dam to that purpose. Engineers located the 10-foot-high, 200-foot-long concrete dam, named after the property owner, near the confluence of Dulzura and Jamula Creeks about two miles upstream from the Otay Reservoir. The intent was to redirect water from the dam via a three-mile conduit for storage in Upper Otay Reservoir. In this 1927 image, city engineers are shown walking the dam's crest investigating for damage the dam may have experienced from a recent flood episode. Note the hillside outpost US cavalry troops used while patrolling the US-Mexican border during World War I. (SDCWHA.)

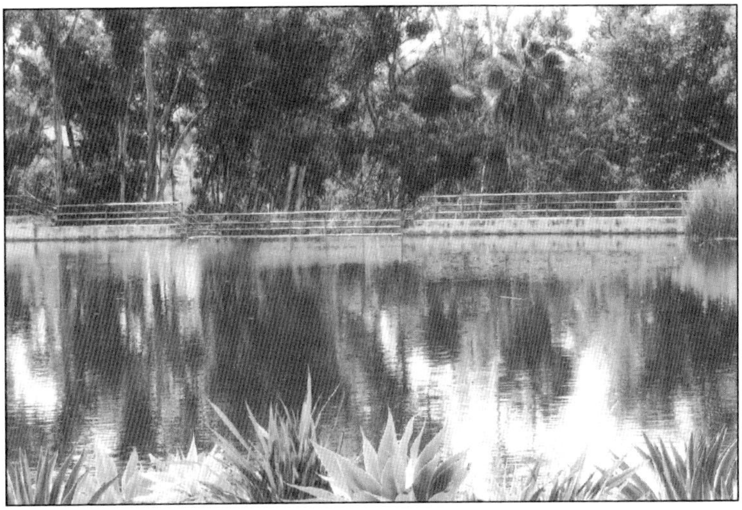

The Fairbanks or Rancho Santa Fe Dam carries little importance beyond its historical value. In the mid-1920s, actor Douglas Fairbanks and his wife, Mary Pickford, purchased a 3,000-acre tract in the old Rancho Santa Fe Land Grant for investment purposes and named it Rancho Zorro after the hero of his popular movie. (Author's collection.)

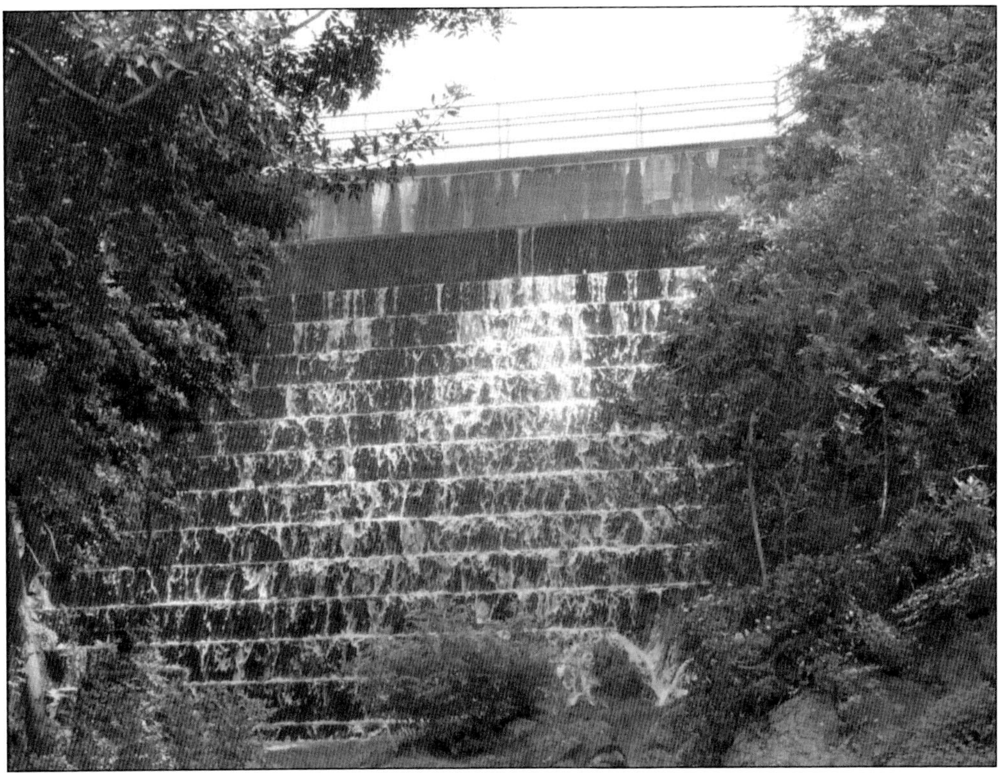

Fairbanks's idea was to plant hundreds of acres of orange trees. To create a water source, in 1926, he constructed a dam to conserve nearby Lusardi Creek. Fairbanks described the dam as a "concrete staircase" because of the 13 descending steps that formed the downstream face. During construction, legend has it that Fairbanks and his wife pressed their hands into the fresh concrete. (Author's collection.)

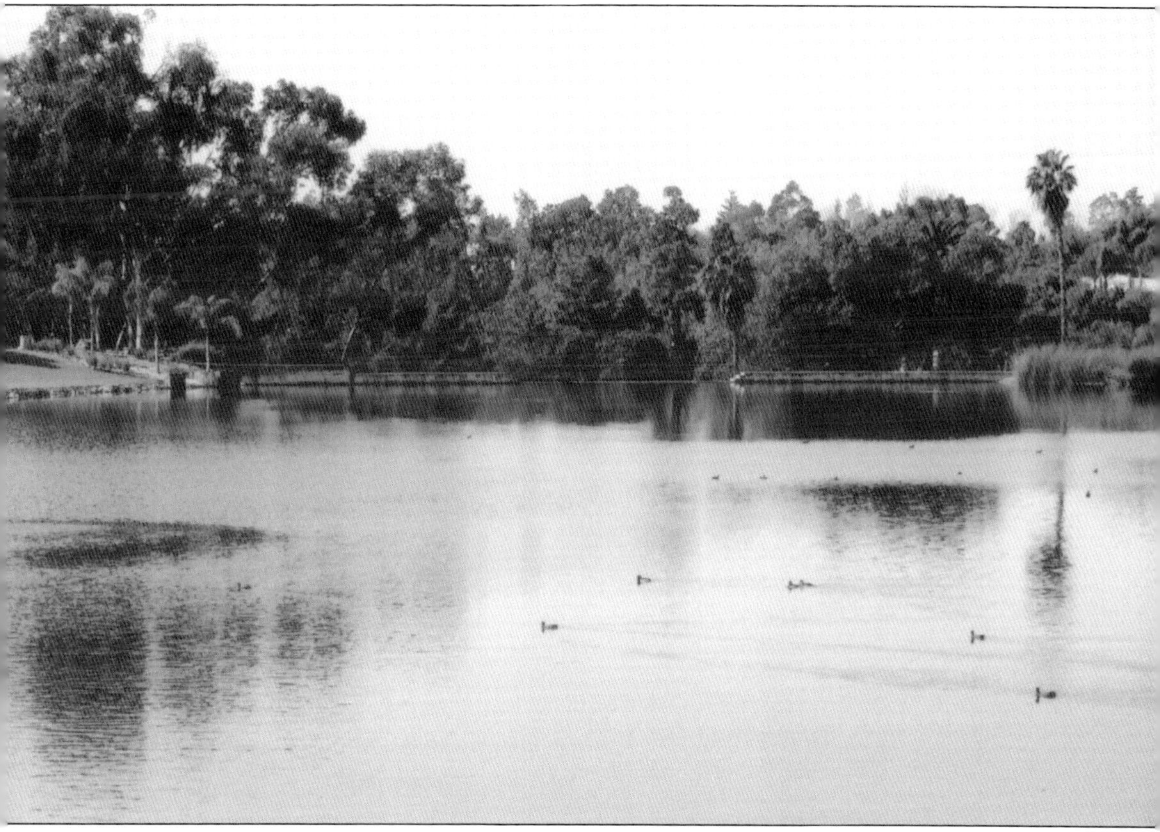

R.E. Struve was the original contractor for the concrete masonry Fairbanks Dam. The structure stands approximately 40 feet high by 155 feet long, with a four-foot crest and a 24-inch release pipe at the base. Water is released via a shallow-sloping two-foot-wide outflow trough that runs horizontally along the upper face. There are two spillways, a 30-foot-long, two-foot-deep notch in the middle of the dam and a 20-inch outfall pipe at the base of the dam. Today, estate homes and a small private community park surround the dam. Through time and natural action, the spread of algae and the increased salinity from groundwater have destroyed the once excellent fishing in the lake. The Fairbanks Ranch Association has initiated plans to restock the lake and open it for fishing to community members by 2018. (Author's collection.)

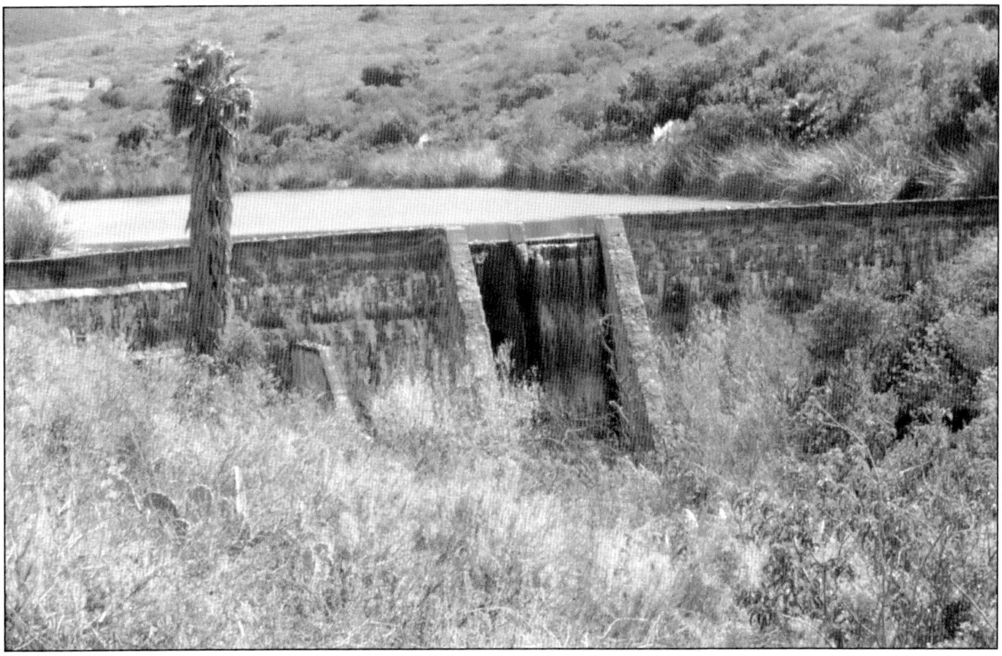

Located near the community of Rancho Penasquitos, the Black Mountain Dam or the Upper 4S Ranch Dam was constructed in a narrow, cactus-studded canyon in 1927 to conserve water for the sprawling 4S Ranch. An Englishman named Joseph Snook settled about 3,000 acres in this area in the 1840s. The name allegedly came from the S for Snook and the four leagues of land the Mexican government granted him. The Ralphs family, of the Ralphs grocery chain, secured the property in the 1980s, maintained 1,200 acres as a working ranch, and designated the remaining lands for development of the unincorporated community of 4S Ranch. (Both, author's collection.)

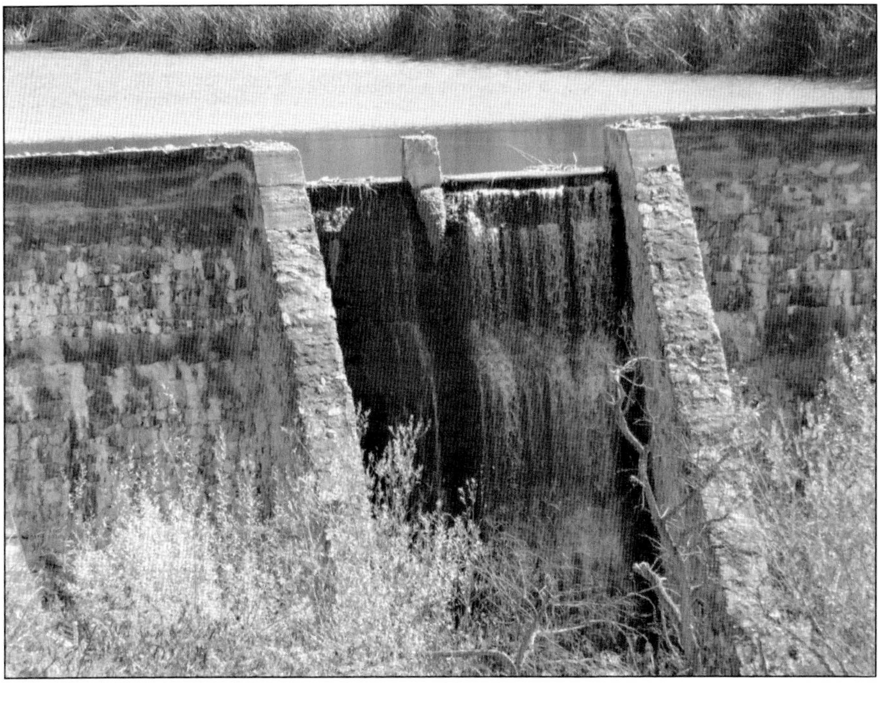

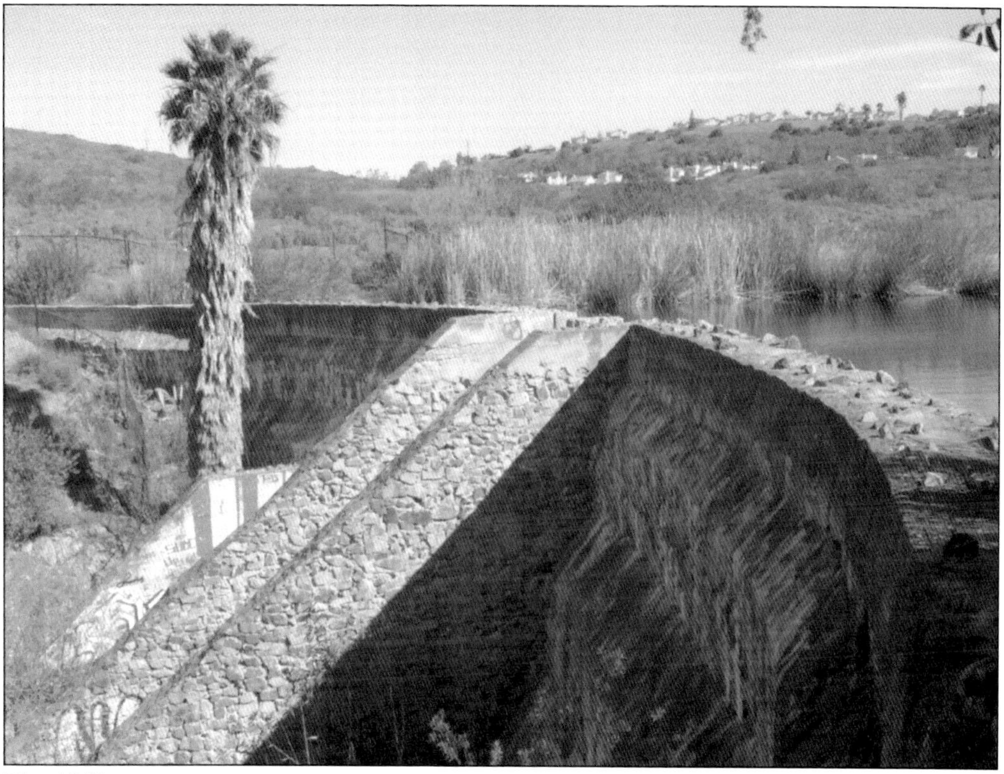

The 4S Dam is a masonry, constant-radius, arch-type dam that conserves Lusardi Creek. It is approximately 240 feet in length and 50 feet high, with a four-foot, riprap-covered crest. Masonry buttresses extend at a 45-degree angle to the streambed, where they are secured to support the structure. The dam's unique spillway is a 20-foot-wide, three-foot-deep notch in the center of the crest. (Author's collection.)

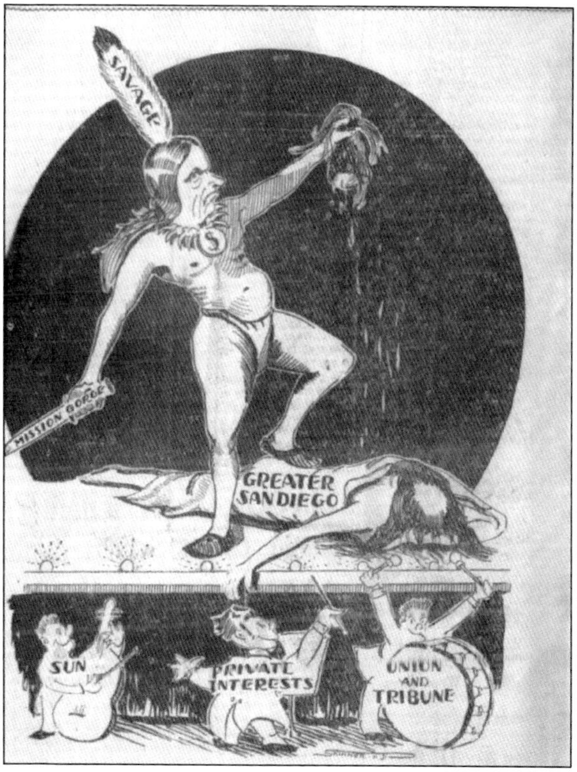

The construction of El Capitan Dam was almost anticlimactic compared to the city's struggle to secure the water rights to the river and decide where to place a river dam. Savage and Fletcher favored a dam in Mission Gorge, while the opposition supported a dam at El Capitan. When voters approved the El Capitan site, an anti-Savage faction accused him of trying to "scalp" the city for personal reasons. (Author's collection.)

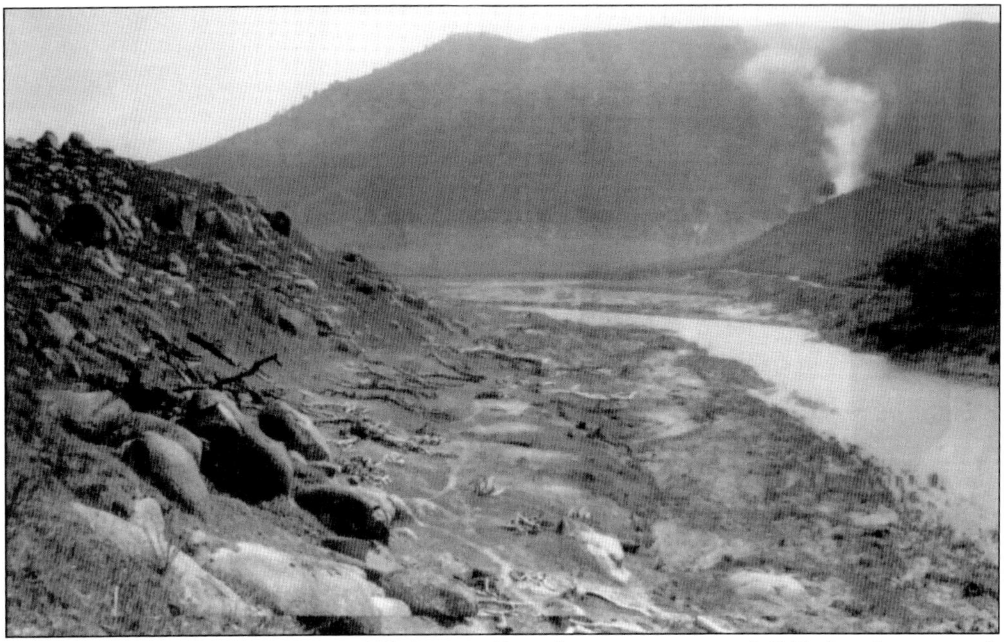

In 1924, San Diego voters approved El Capitan Dam, but it was not until 1931 that the city took possession of the site from the La Mesa, Lemon Grove, and Spring Valley Irrigation District and the project went forward. Savage and his engineers selected a site on the edge of the Cuyamaca Mountains 30 miles northeast of the town in a confined canyon where the San Diego River departed the foothills. Savage favored a concrete gravity dam, but explorations of the bedrock and the presence of abundant fill and quarry material suggested that a semi-hydraulic rock-filled embankment structure was more appropriate. The city constructed El Capitan Dam between 1932 and 1934 at a cost of $5.8 million. (Both, SDCL.)

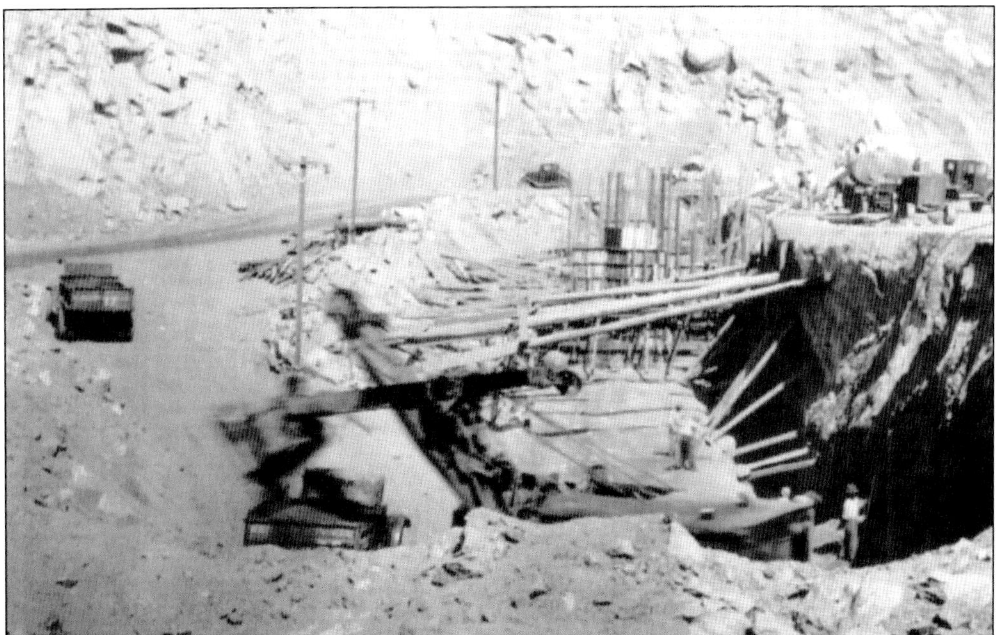

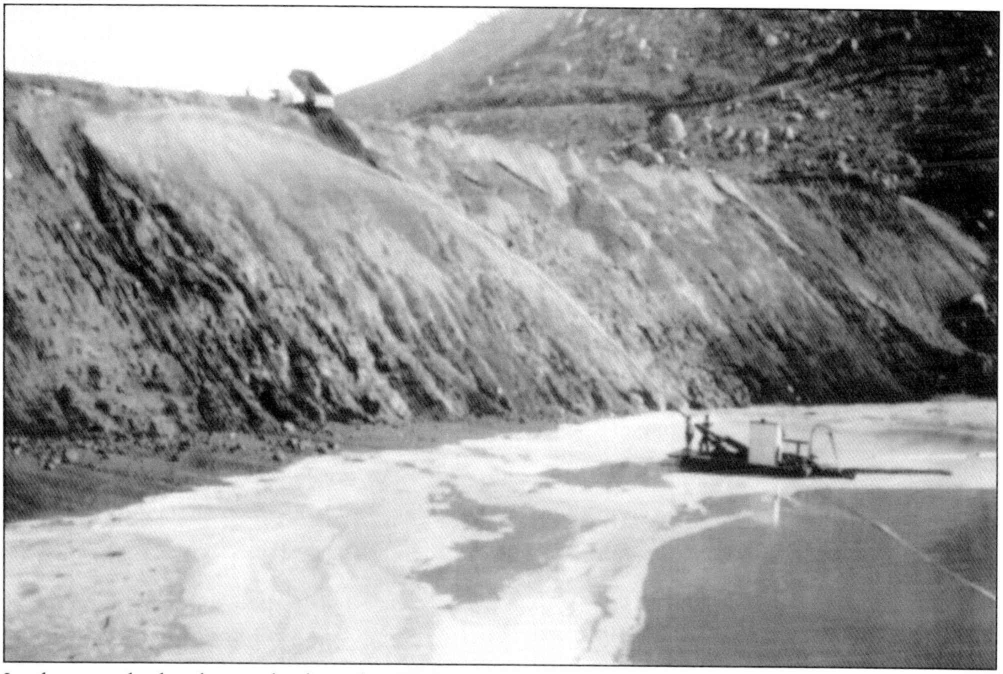

In the semi-hydraulic method used at El Capitan, workers sprayed, or sluiced, a water-and-earth mixture from a borrow pit through a high-pressure water nozzle onto the exterior walls of the dam. This caused a natural sorting of the material: the finer settled into the surface and the coarser remained on the top and congealed. At El Capitan, the workers created a large pool on either side of the dam as the source for the water mixture and sprayed the sluice on the face of the dam from a nozzle mounted on a barge. The workers sluiced the face to a depth of about 17 feet. (Both, SDCWHA.)

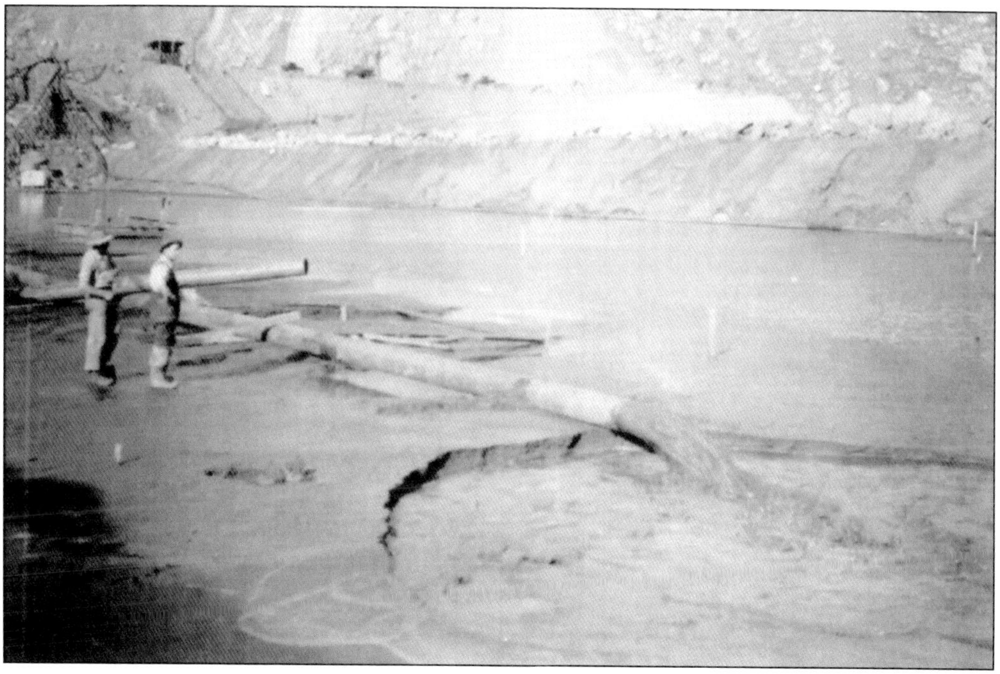

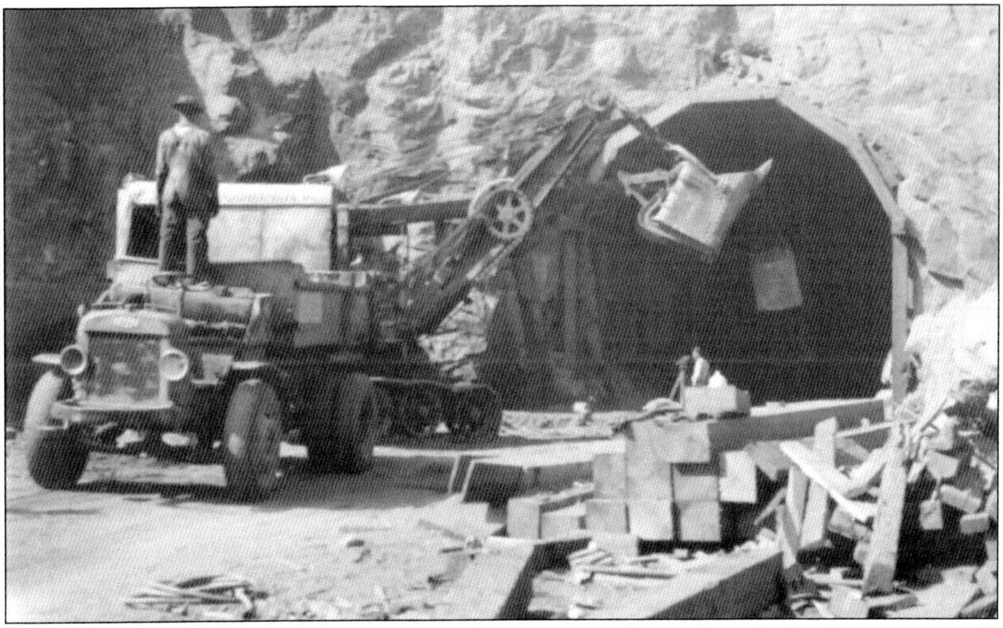

At El Capitan, three concrete-lined, 26-foot-diameter tunnels ran through the downstream wall to create drainage and divert the river while workers infused the interior of the embankment with an impervious clay core consisting of 55,000 cubic feet of material. Water released through the control tower poured into the distribution pipeline through six 30-inch valves. El Capitan featured a concrete overflow channel spillway on its north side that drained when the water level in the reservoir passed the conservation point. The reservoir extended so far into the catchment that El Capitan did not act as a true flood control structure for communities in the lower sections of the river. (Both, SDCWHA.)

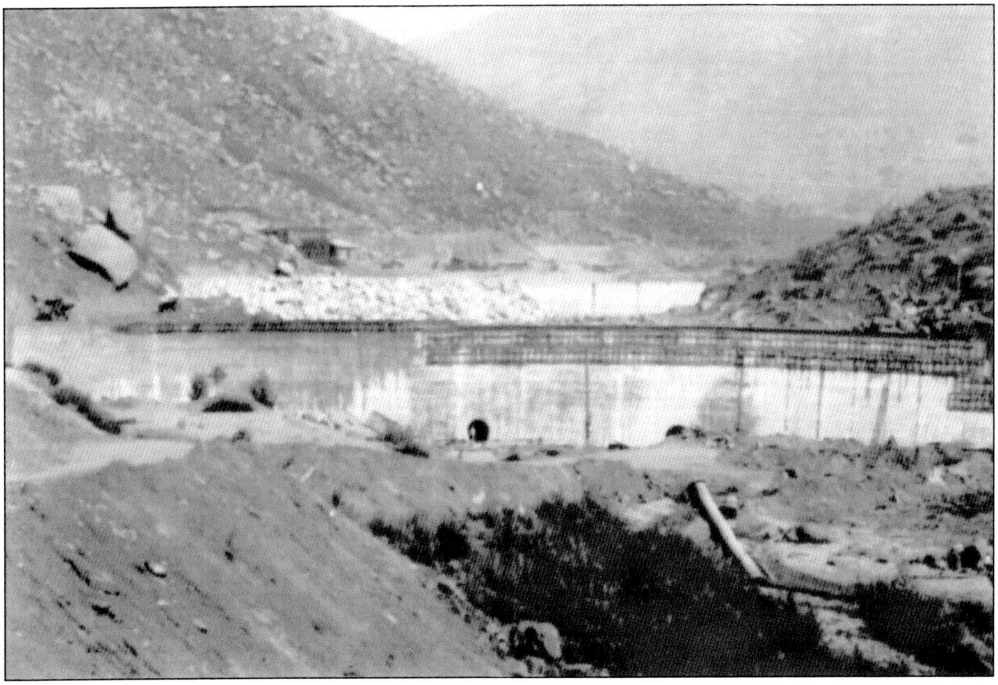

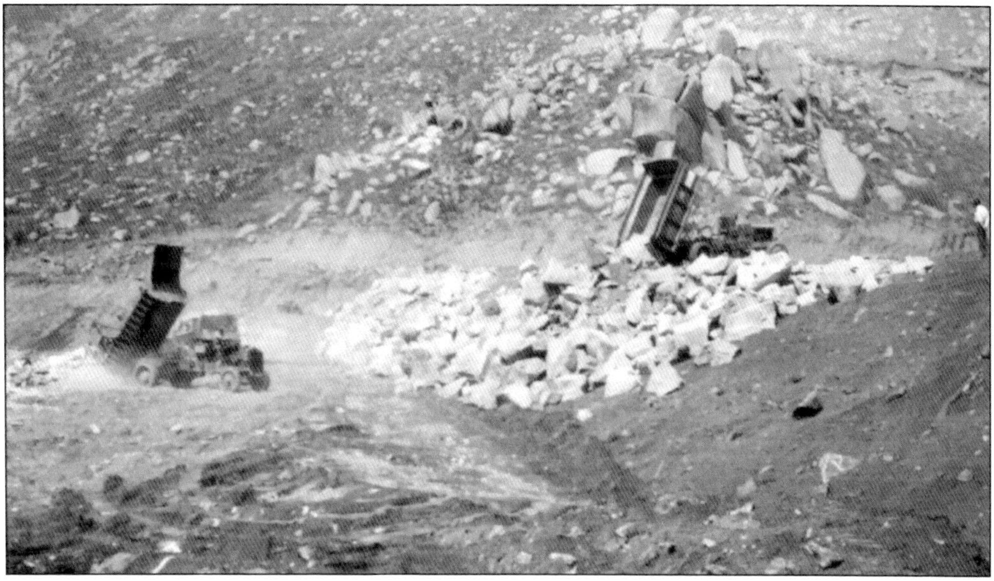

El Capitan did not proceed without legal difficulties. After a bout of design and contractor disputes, Savage recommended H.W. Roehl and T.E. Connelly as the contractors. Immediately, a rival San Diego contractor, Daley and Fenton, brought suit against the city arguing the bidding plans were insufficient for drawing construction plans. Judge C.N. Andrews rejected the injunction, and the city began work. The dam had a crest length of 1,170 feet and stood 217 feet above the streambed. Savage and his fellow engineers anticipated that the completed structure would create a 1,580-acre reservoir conserving approximately 112,800 acre-feet or 39 billion gallons of water. The associated pipeline would have the capacity to supply the city 10 million gallons per day. The reservoir first filled to capacity during the storms of 1938. (Both, SDCL.)

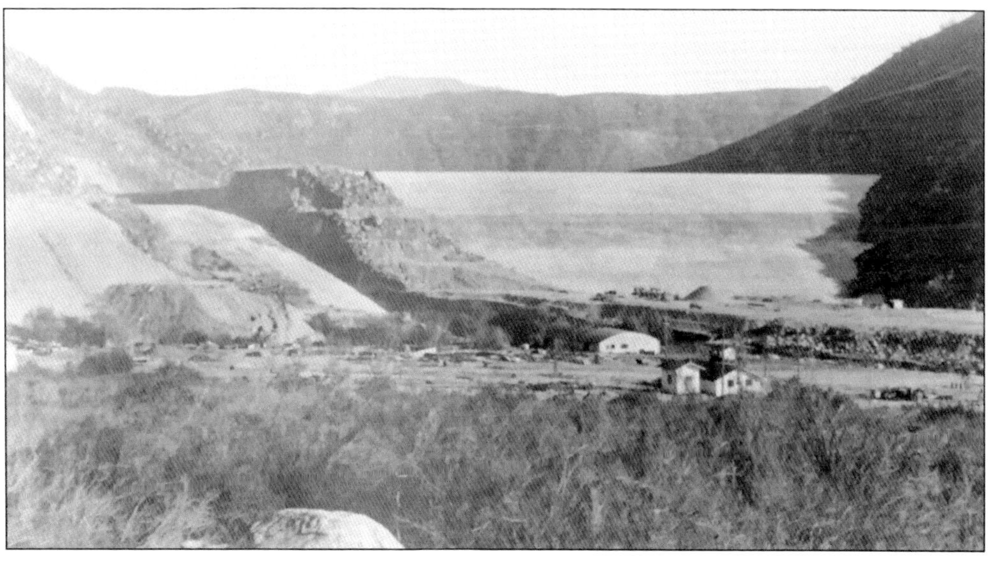

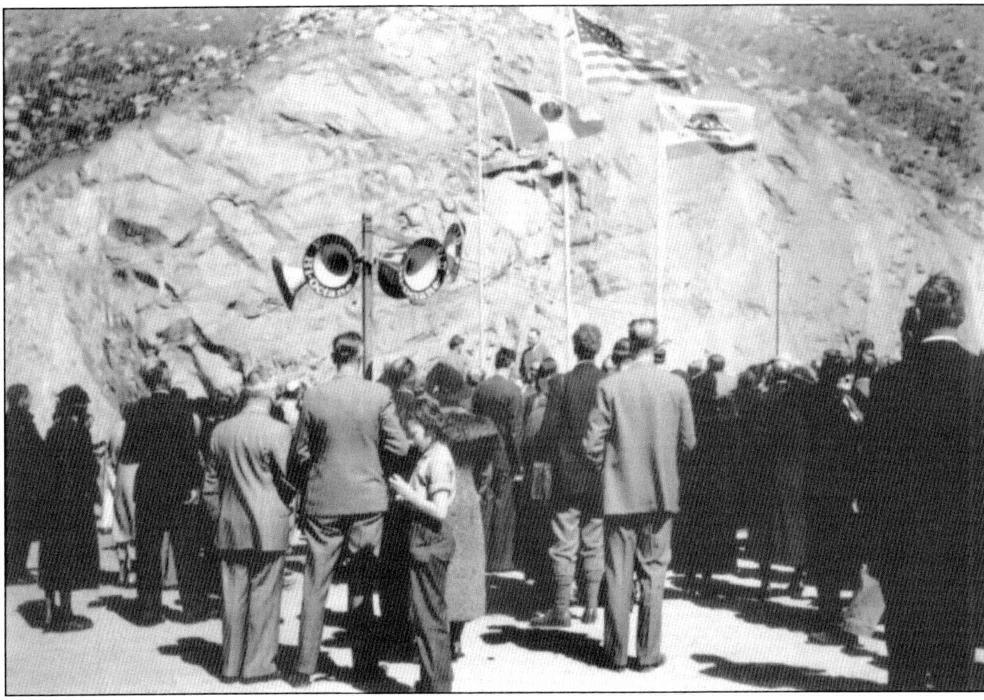

The city dedicated El Capitan Dam on February 22, 1935. Ironically, the dedication plaque bears the names of an acting mayor, a new city manager, and several people who actively opposed the project, but no names of the individuals who made the dam possible. When completed in 1934, the dam created the city's largest reservoir. Unfortunately, Savage died before the dam was completed. The city paid homage to its greatest dam builder and renamed Lower Otay Dam Savage Dam in his honor. But El Capitan's success was transitory, as the city's growing population re-engaged the quest for water from the Colorado River to sate its domestic needs. (Both, SDCWHA.)

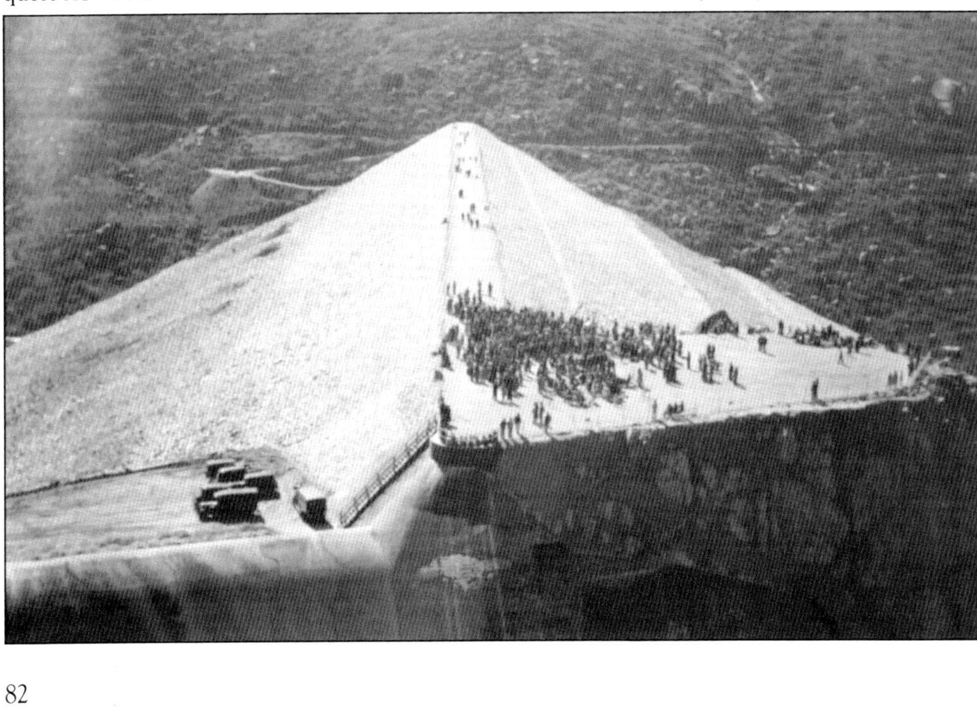

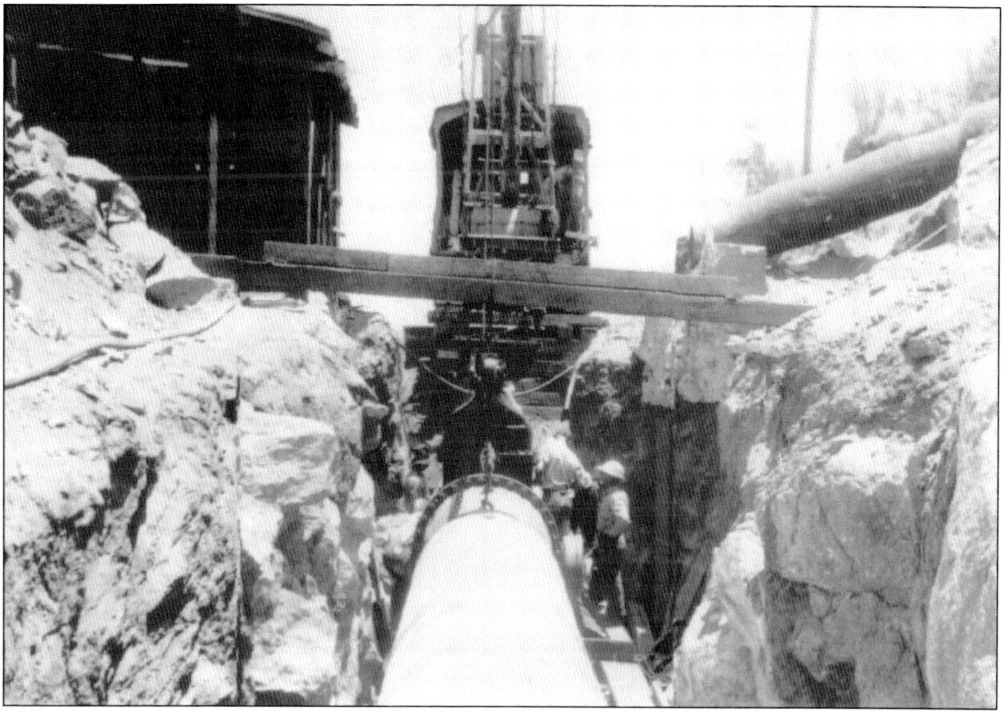

The city built the El Capitan Pipeline to transport the water to the city distributing system. The line was completed in 1934 along with the dam. A 48-inch pipeline ran from the dam to the Lakeside Pump Station, then via the El Monte pipeline to the Alvarado treatment plant below Murray Reservoir for city consumption. The completion of the dam ended the service of the old flume system. (SDCWHA.)

While the El Capitan project temporarily relieved the city's water dilemma, it disrupted the lives of the Indians living on the lands the reservoir flooded. Despite protests, the Department of the Interior reimbursed and relocated the El Capitan Grande Indians to adjacent properties of their choice. At the time, the fate of the Native Americans was tangential to the city's water needs. (SDCL.)

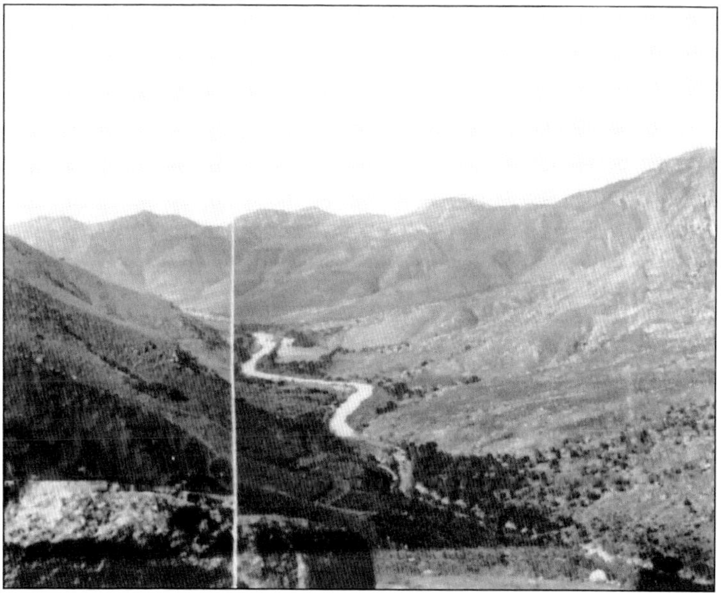

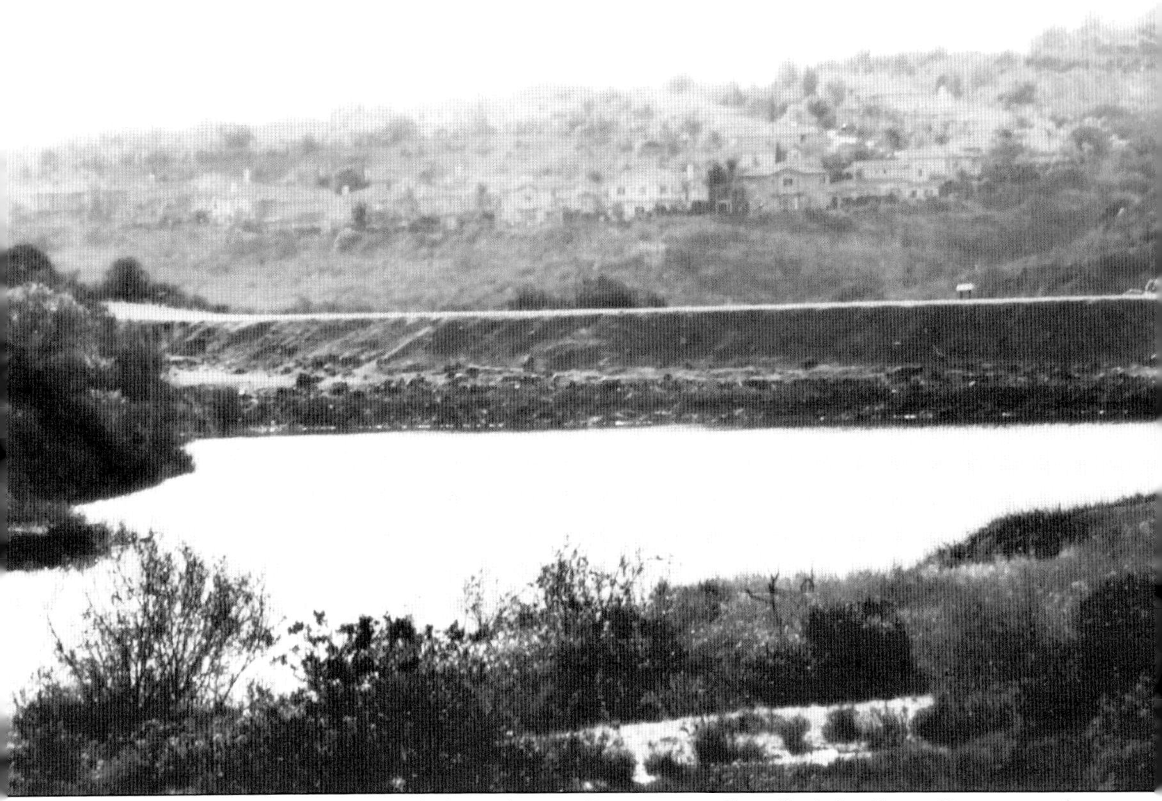

The City of Carlsbad designed the Calaveras Dam, originally called the Fraser Dam, to end its reliance on pumped river water. In 1910, Carlsbad received the majority of its water from the supply the Oceanside Mutual Water Company piped in from wells on the San Luis Rey River about three miles north of Oceanside. In 1914, the South Coast Land Company created the Carlsbad Mutual Water Company to develop Carlsbad's water system, again using a system of river wells. The company could pump water as long as the water table did not drop below the level the state Division of Water Resources had set. Nevertheless, the drilling depleted the water level to the point where seawater was leaching in and contaminating the river system. Within 20 years, the town desperately needed an alternative water source, and the construction of a dam and reservoir seemed the answer. (Author's collection.)

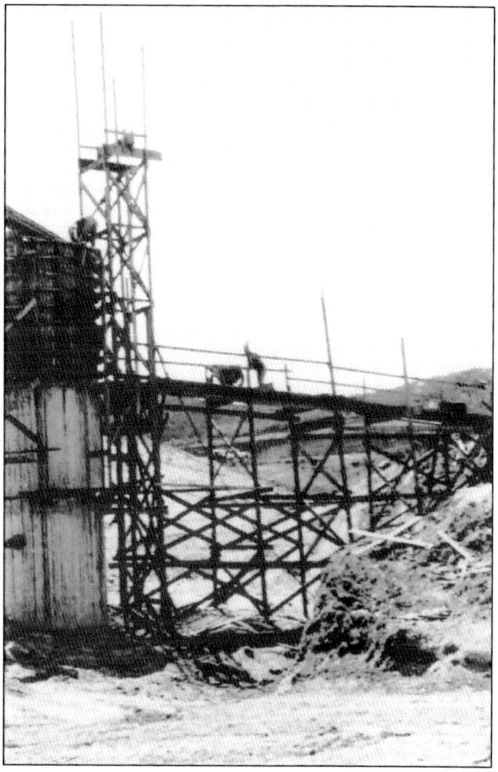

In the summer of 1941, the City of Carlsbad constructed the 67-foot-high, 490-foot-long, earth-filled Calaveras Dam with 85,000 cubic yards of material taken from the contiguous hillside. The water officials named the dam after the creek it straddled. The city built the dam to create a water supply reservoir that would accept natural runoff from the Calaveras Basin, 150 acre-feet of water per year diverted from Calaveras Creek, and water pumped from the wells on the Lower San Luis Rey River. An extended drought and the saltwater intrusion into the river system compelled the Carlsbad Mutual Water Company to cease using well water by 1949. (Both, courtesy of Carlsbad History Collection, Carlsbad City Library.)

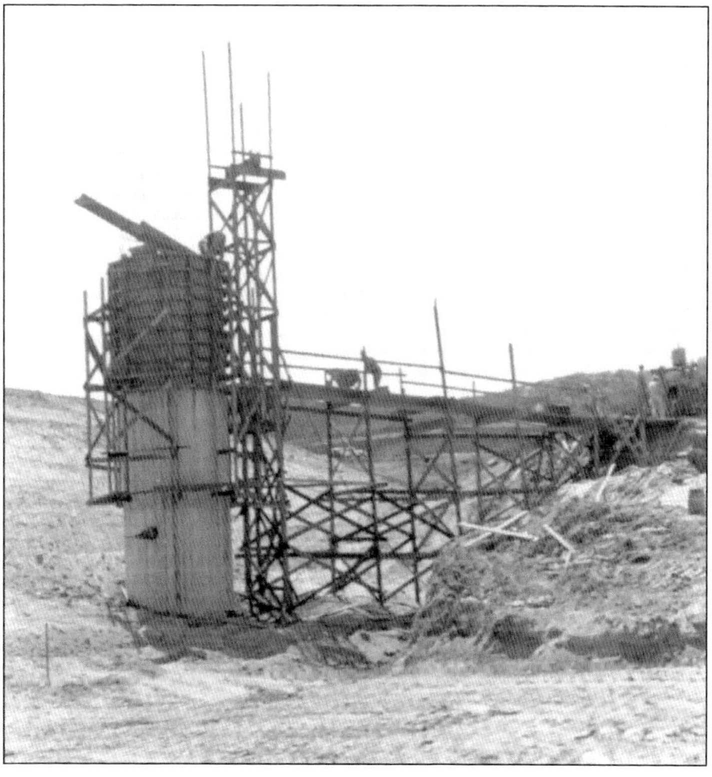

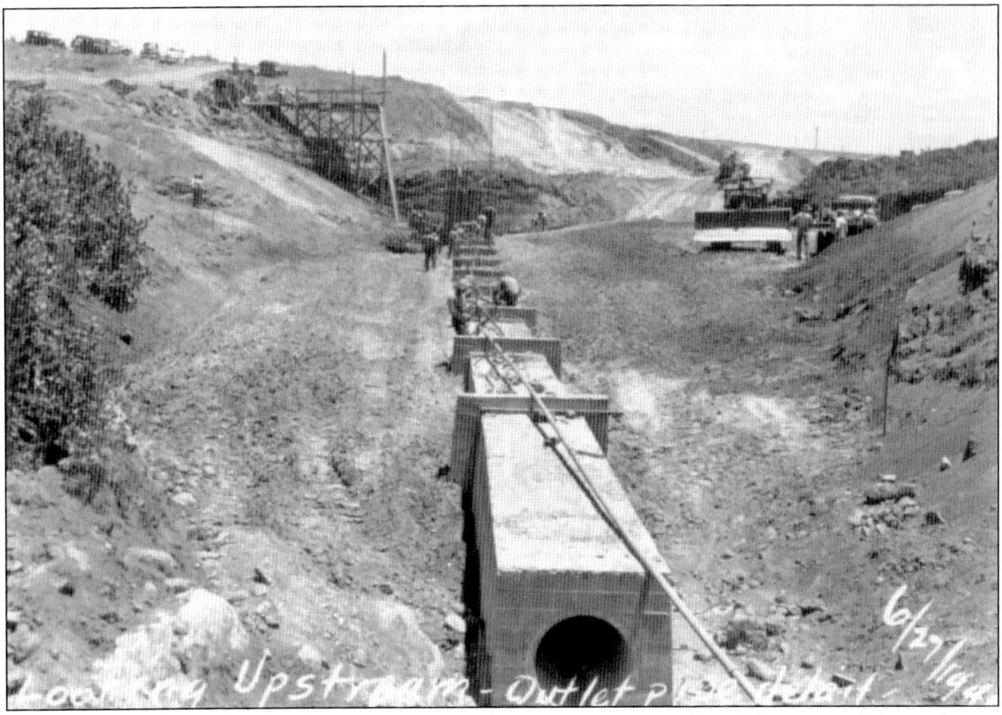

Calaveras's 520-acre-foot-capacity reservoir outlets water via a pipeline. Engineers did not design the dam as a flood control structure, but they did size the spillway, a simple overflow on the western end of the dam, to protect against a protracted flood event. Interestingly, the reservoir is situated beneath a 22-million-year-old extinct volcanic plug. In the early 1900s, locals stripped away the west face of the plug and mined the area for gravel, a practice that continued into the 1930s. The Carlsbad Municipal Water District manages the Calaveras Dam and Reservoir, part of the 400-acre Lake Calaveras Reservoir nature preserve, which features miles of biking and hiking trails, one of which crosses the crest of the dam. (Both, courtesy of Carlsbad History Collection, Carlsbad City Library.)

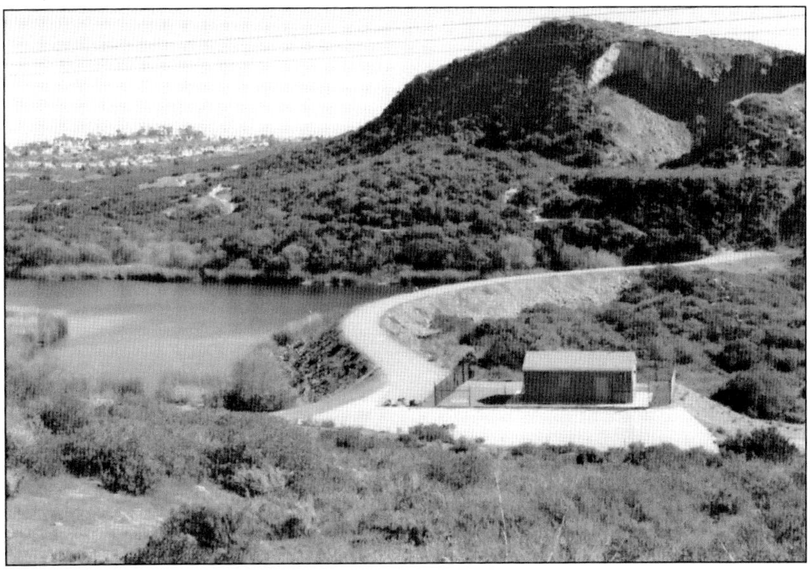

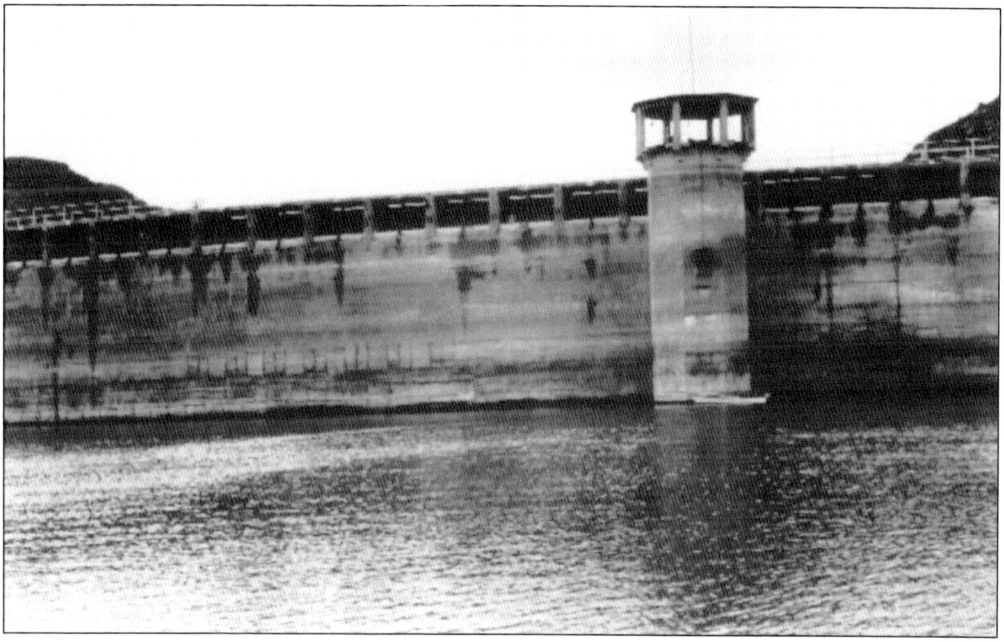

San Diego water pundits had considered a dam on the San Vincente Creek since the 1920s. The possibility of acquiring Colorado River water revitalized the project in 1939, but under the malaise of the Great Depression, city voters rejected the bond election. The city's rising population and pervasive need for additional water forced the electorate to approve the project a year later. (SDCWHA.)

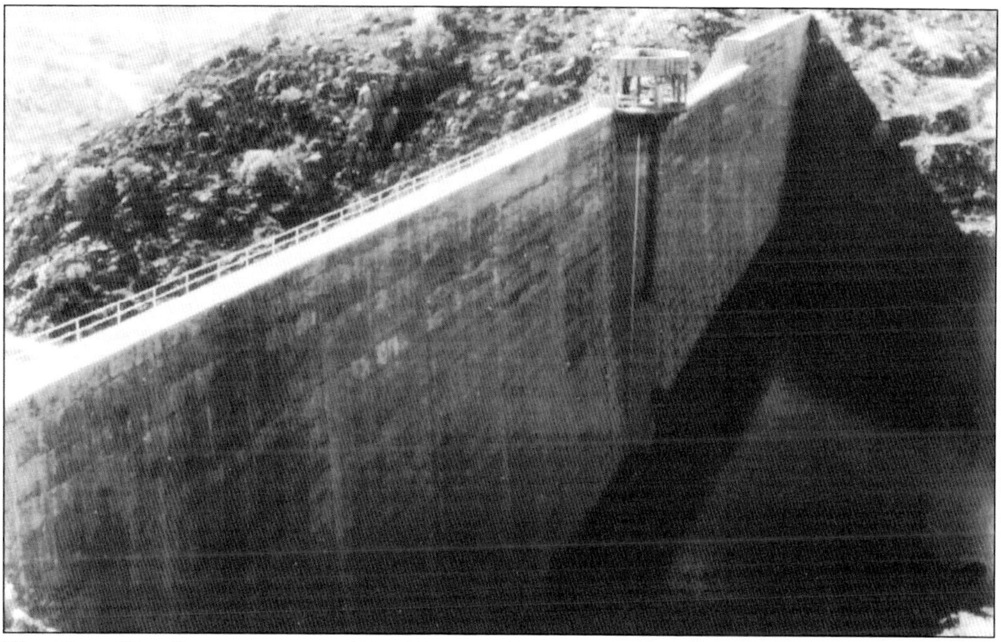

The City of San Diego constructed San Vincente Dam in a canyon about 15 miles northeast of the city between 1941 and 1943. The dam was efficiently located to serve 85 percent of the city through a gravity flow system and also receive water piped from the El Capitan and Sutherland Reservoirs and from the projected imports of Colorado River water via the San Diego Aqueduct. (SDCWHA.)

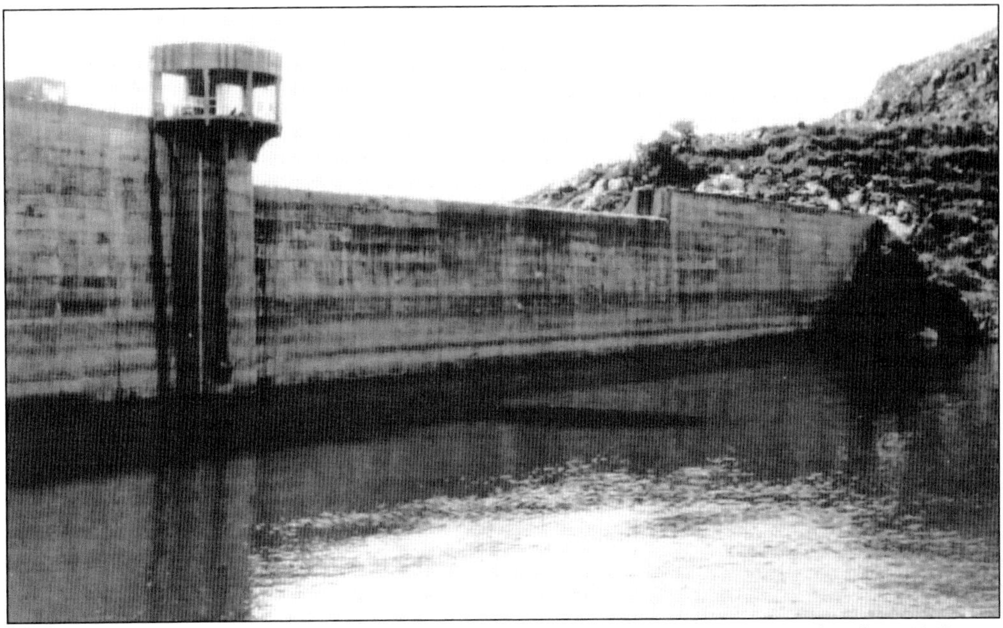

The original San Vincente Dam was a gravity mass concrete structure completed in 1947 at a cost of $2,767,021. It stood 199 feet high, was 980 feet in length, and conserved a 116,900-acre-foot reservoir. Five-foot-by-fifty-foot un-grouted concrete blocks covered the face of the original dam, which allowed water to leak into the interior. The spillway was an uncontrolled flow type in the center of the dam. (SDCWHA.)

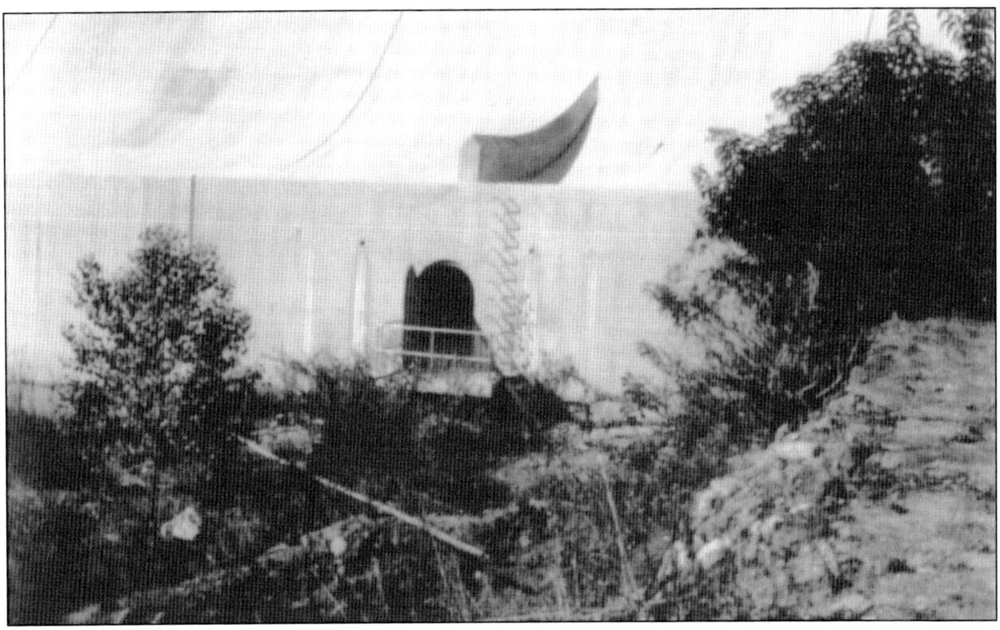

A unique feature of San Vincente was the placement of a 22-foot-wide tunnel passing through the base of the structure to accommodate an existing county road. With the dam's completion, that tunnel and a small diversion tunnel were concreted closed. When filled, the reservoir, designed for storage, flood control, and recreation, submerged several Indian campsites, a historic train depot, and a proposed railroad route. (SDCWHA.)

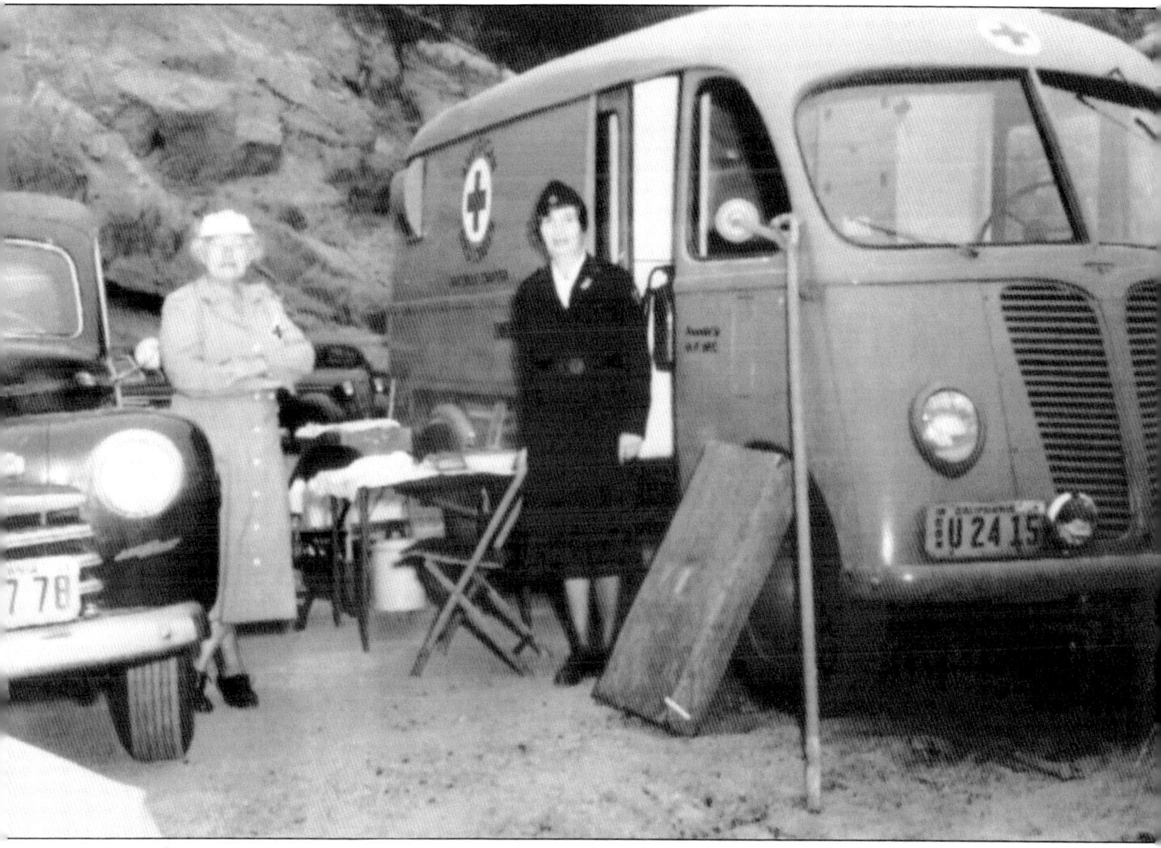

Opening day at San Vincente in 1947 was a spectacular civic event celebrating the completion of the project and the beginning of the possibility of bringing Colorado River water to the city. Briefly, it appeared that the government might renege on the deal to bring the water through a new San Diego Aqueduct, but the civic uproar and the growing Navy presence convinced Congress to fund the project. Local newspapers announced the festivities, and citizens arrived in record numbers. The stocked lake was open to local fishermen, and the local Red Cross chapter was on hand to deal with sunburn and any minor accidents. The City of San Diego and the San Diego County Water Authority jointly administer the facility. (SDCWHA.)

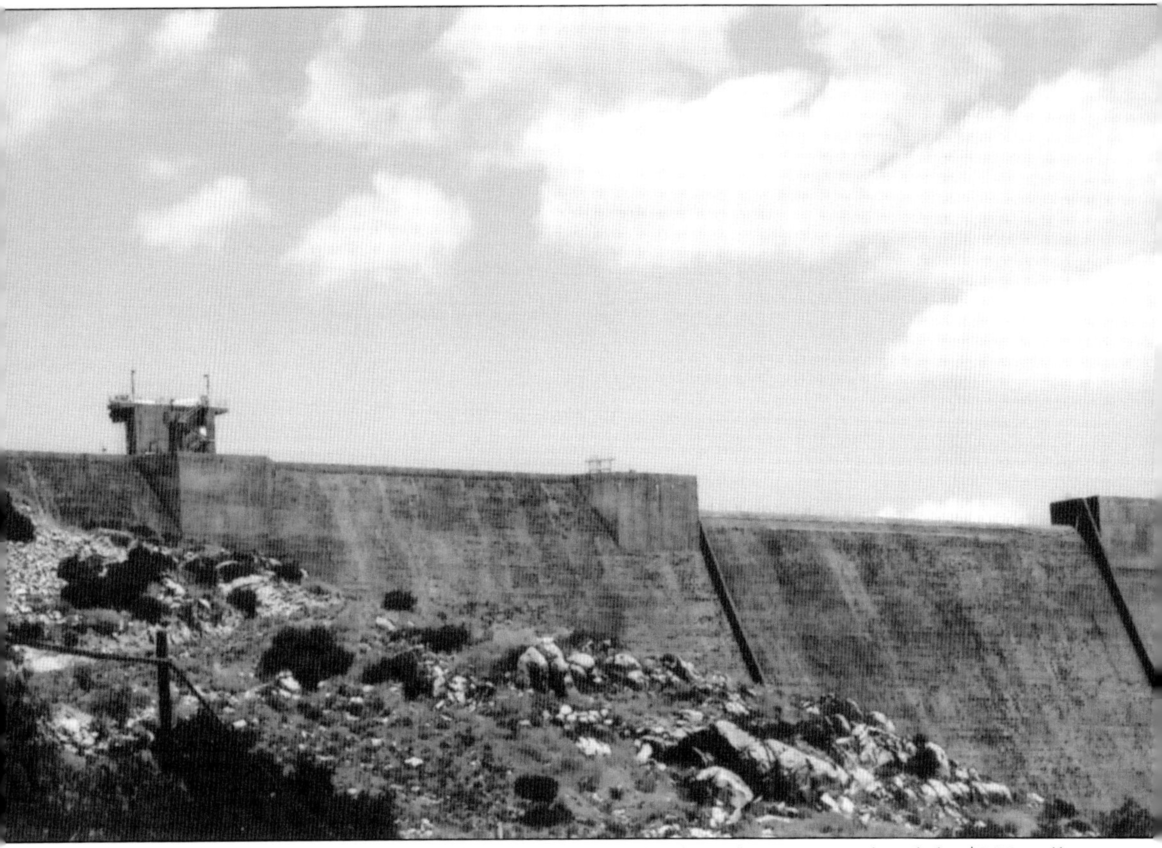

San Vincente Dam gained national attention in 2016, when the city completed the $838 million project to raise the height of the dam an additional 117 feet to 337 feet. As Maureen Stapleton, the general manager of the county water authority, said, "we super-sized" the dam. Using the rolled compacted concrete method, this became the largest rise of any dam in the United States and the tallest roller compacted dam in the world. The reservoir is connected to the two aqueducts that receive water from the Metropolitan Water District and the Imperial Valley Irrigation District. The work doubled the reservoir storage capacity to 226,000 acre-feet, making it the largest reservoir by volume in the county. Opening day after an eight-year closure was reminiscent of the 1947 event, as huge numbers of San Diegans turned out to celebrate, but in this case, many of the anglers and boaters complained of the higher rental and use fees. (SDCWHA.)

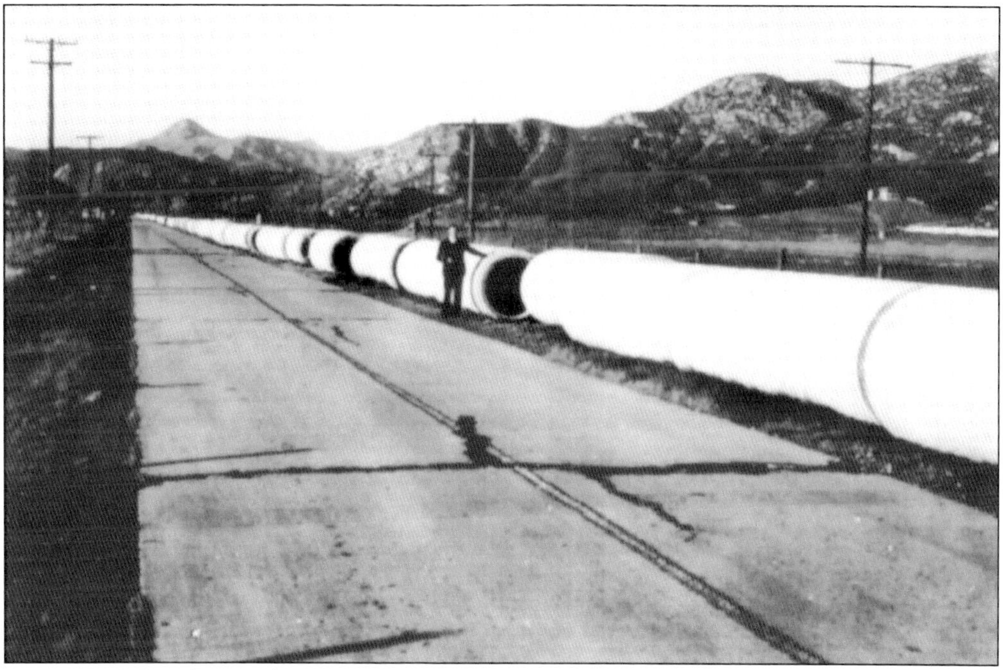

Between 1945 and 1947, the US Navy constructed a pipeline the Bureau of Reclamation designed to transport Colorado River and San Diego Aqueduct water to San Diego. The Navy became involved in the $17.5-million project because of the need for water to support the massive housing boom World War II ignited in San Diego. The first precast concrete pipeline ran 71 miles from San Jacinto to San Vincente Reservoir. (SDCWHA.)

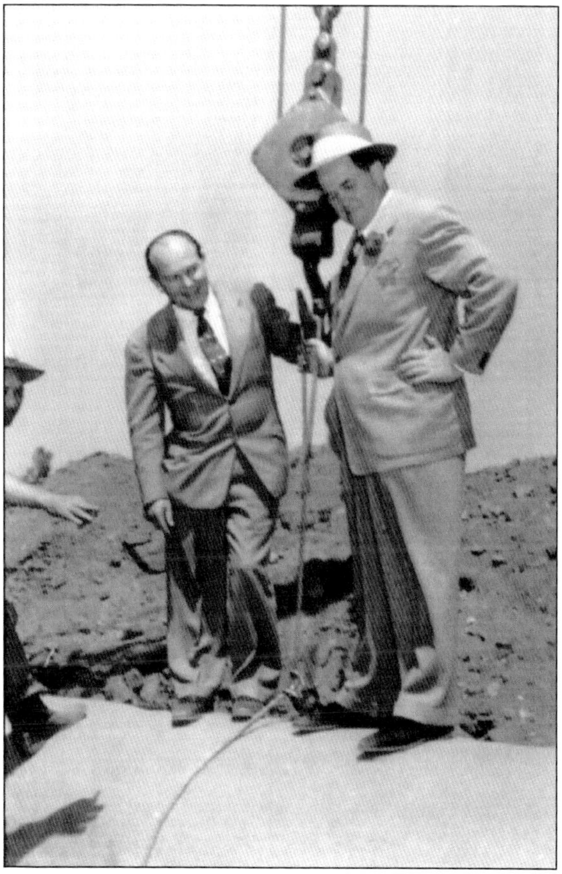

A second aqueduct, with pipe ranging from 48 inches to 96 inches in diameter, was completed between 1952 and 1957. These lines moved water from the Colorado Aqueduct to Savage Reservoir. Throughout the project, San Diego mayor Percy Benbough made periodic visits to generate public support and mark the progress of the pipeline toward San Diego. (SDCWHA.)

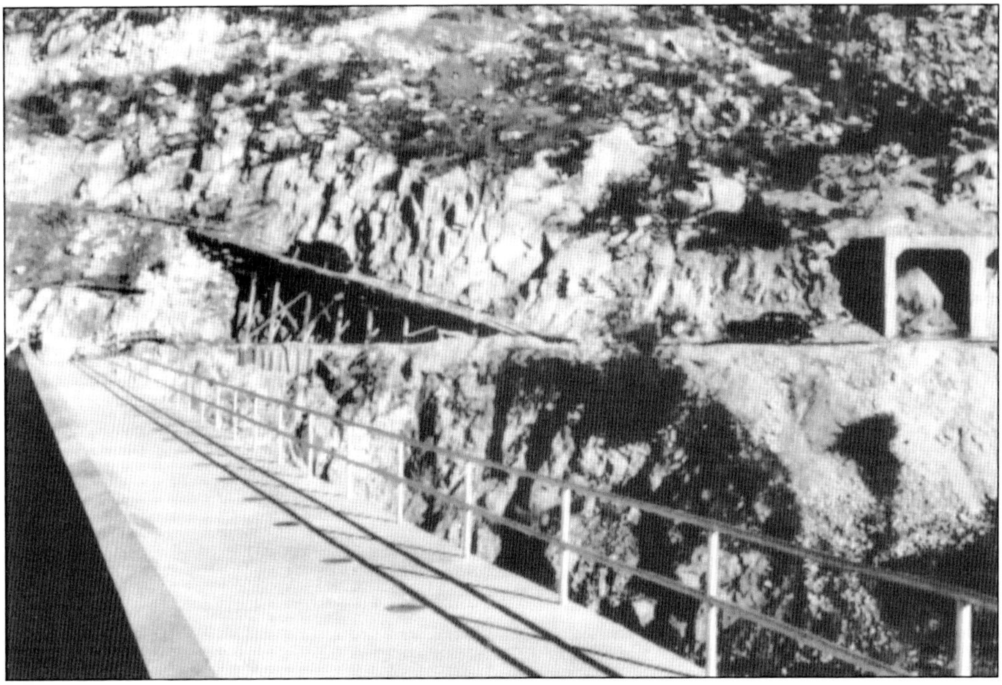

The government almost abandoned the aqueduct. With the war ending, the Navy cancelled the project, but the persistence of the San Diegans saved it. Approximately 850 laborers worked on the aqueduct, with unskilled workers earning 87¢ per hour while skilled workers made $1.75. The Colorado water was transferred via the Colorado Aqueduct to Lake Skinner in Riverside, where the San Diego Aqueduct carried it to San Vincente. (SDCWHA.)

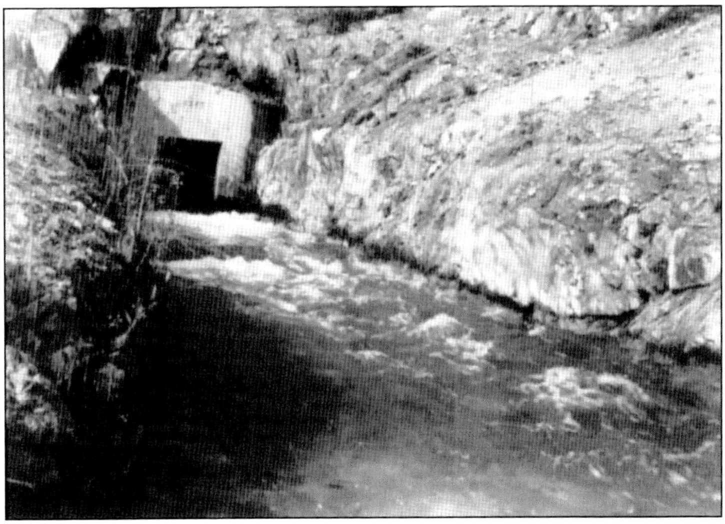

The first aqueduct water arrived in November 1947. The water entered the San Vincente Reservoir immediately behind the west side of the dam's crest. A second aqueduct line was added later, running from Diamond Valley Lake, which collected water from Northern California and sent it to the Sweetwater and Savage Reservoirs. (SDCWHA.)

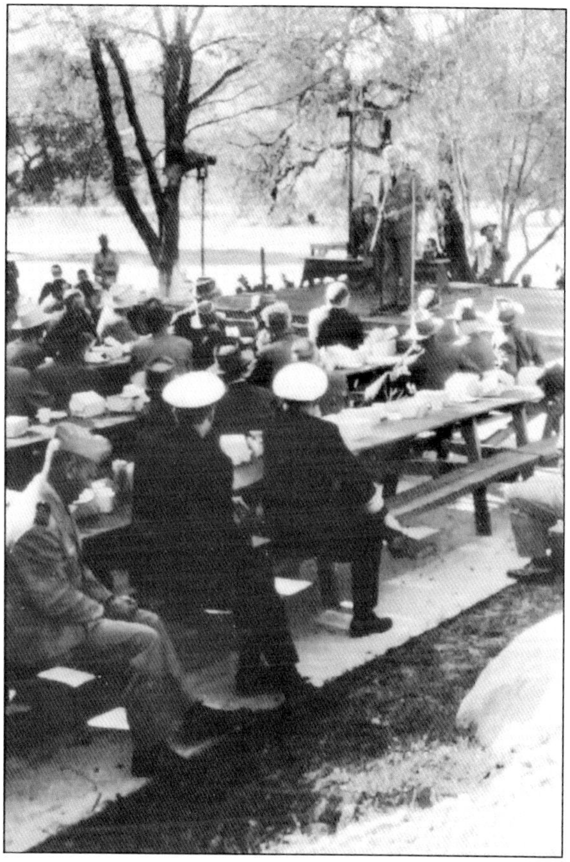

The dedication ceremony for the aqueduct took place in December 1947 in Foster Park, just below San Vincente Dam. The main speaker, Navy captain Alden K. Fogg, noted that the project went forward because Pres. Franklin D. Roosevelt, a former undersecretary of the Navy, understood the importance of the aqueduct to the Navy and to San Diego. To expedite the project, FDR had personally cut through bureaucratic red tape and ordered the Navy to get water into San Diego County to support the government's present and future military assets and the ballooning city and county population the war industries had created. A stable water supply was essential to support the city and the Navy. (Both, SDCWHA.)

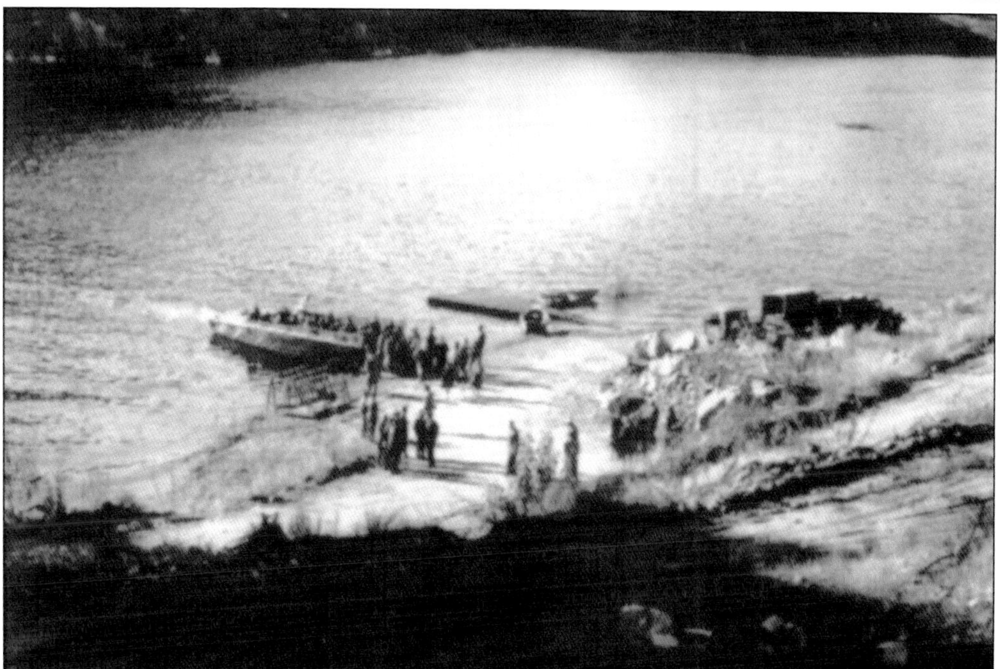

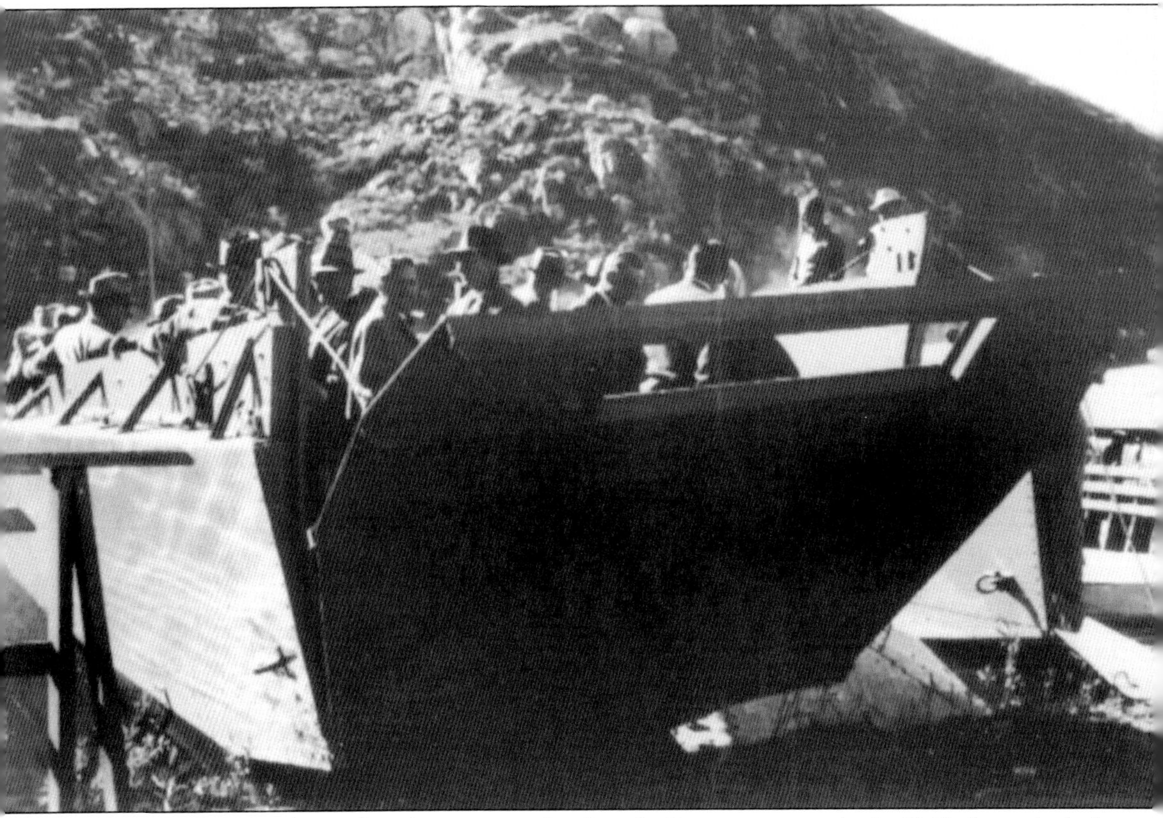

For a project that began less than two weeks after the Japanese surrender in 1945, the arrival of the first aqueduct water in November 1947 was timely. At that point, San Diego County had only a three-week supply on hand. After a brief fight, the Metropolitan Water District allowed the water to flow through its lines into the aqueduct. To celebrate the arrival of the aqueduct water, the Navy brought in old World War II landing craft, loaded guests and dignitaries aboard, and ferried them around San Vincente Reservoir while guides pointed out the features of the reservoir and the aqueduct. The completion of the aqueduct signaled the end of San Diego's mission to engineer the town's way out of the water dilemma and the beginning of the city's water addiction and its dependence on imported water. The county constructed additional lines in 1954, 1960, and 1973. Today, about 90 percent of the county's water is transmitted through the four active aqueducts. (SDCWHA.)

Floods in 1916 and 1927 motivated the Sweetwater Authority to protect and collect domestic water for the towns of Chula Vista, National City, and Bonita. As a result, the California Water and Telephone Company constructed Loveland Dam across the upper Sweetwater River in 1945. Most locals called the structure the Sweetwater Falls Dam, but it was officially named after the company president, Chester H. Loveland. (Courtesy of Bonita Museum and Cultural Center.)

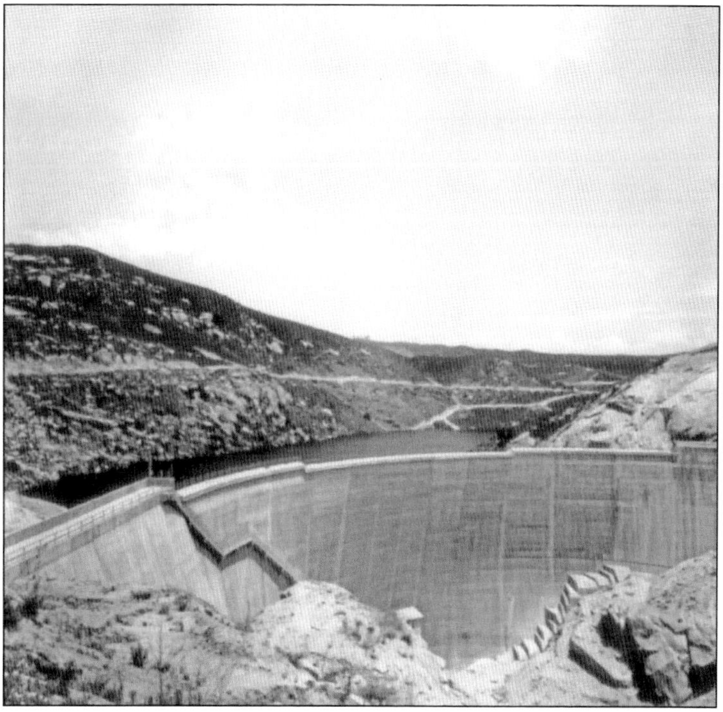

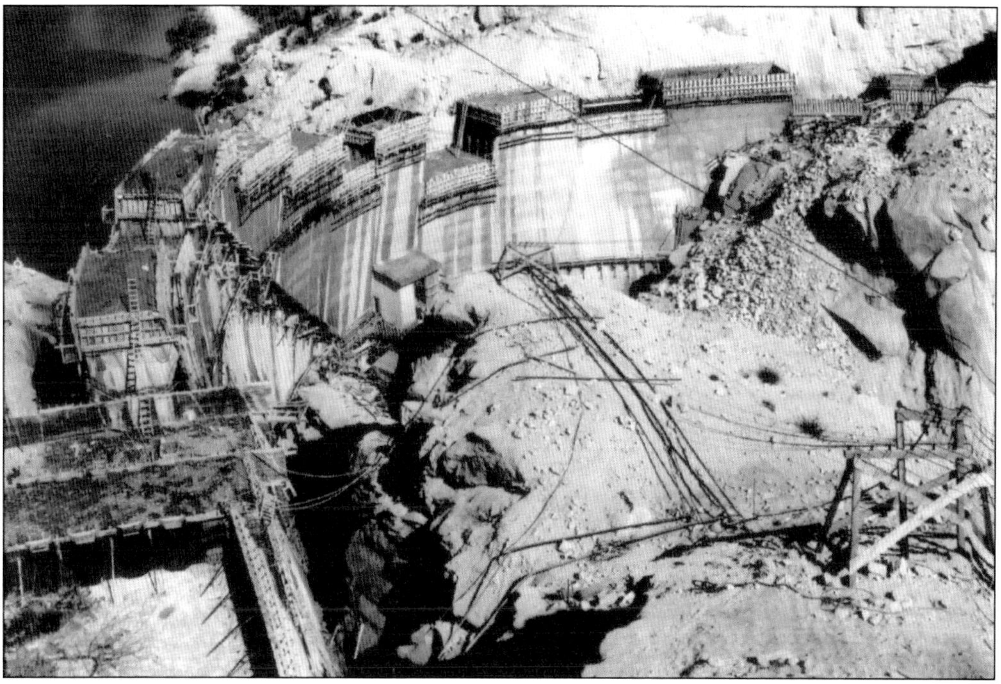

Located 34 miles from San Diego, the 203-foot-high, 765-foot-long concrete thin-arch dam has a simple over-the-crest spillway in its middle. The structure took shape as workers layered vertical rows of huge individual concrete blocks. As the concrete in the lower block cured, workers constructed a wooden frame atop it for the next concrete pour. (Courtesy of Bonita Museum and Cultural Center.)

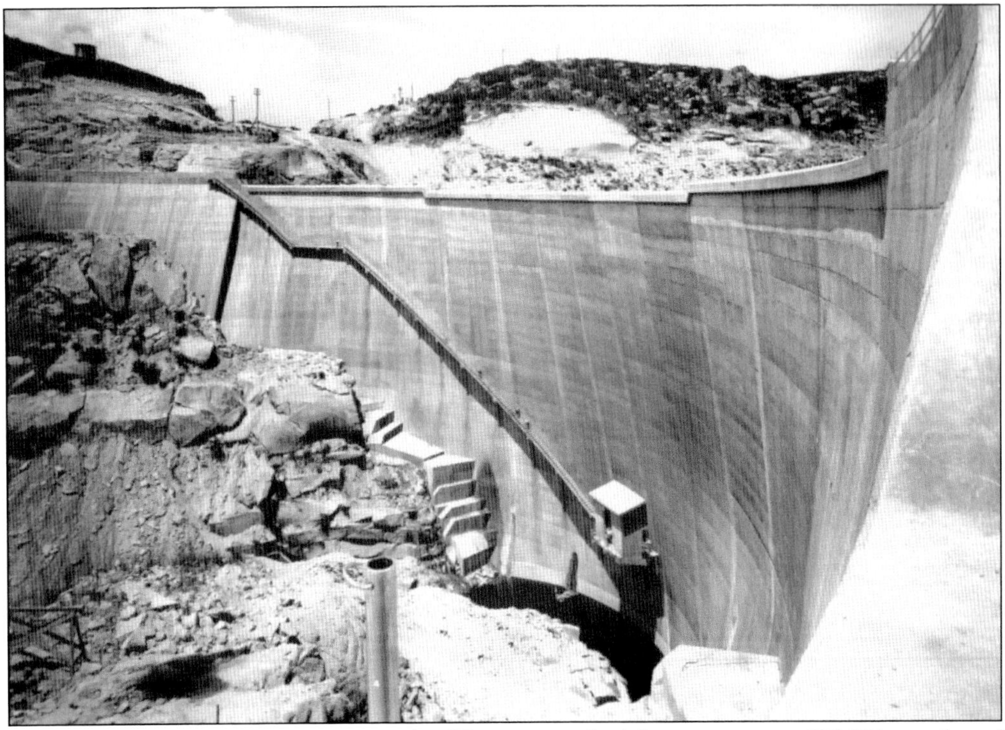

The dam created the long, narrow Loveland Reservoir, which has a capacity of 25,387 acre-feet of water and a surface area of 454 acres. Loveland Reservoir receives natural runoff from the upper reaches of the Sweetwater River, but with a minimal catchment, it is essentially a flood control and storage facility. The bulk of the reservoir water is transferred by pipeline from the El Capitan and San Vincente Reservoirs. The authority can release the collected water southward down the Sweetwater River to the Sweetwater Reservoir when deemed necessary. The US Forest Service transferred the surrounding chaparral- and chamise-covered lands to the authority for public use in 1966. The Sweetwater Authority started water utility operations in 1977 and now manages the dam and reservoir. (Both, courtesy of Bonita Museum and Cultural Center.)

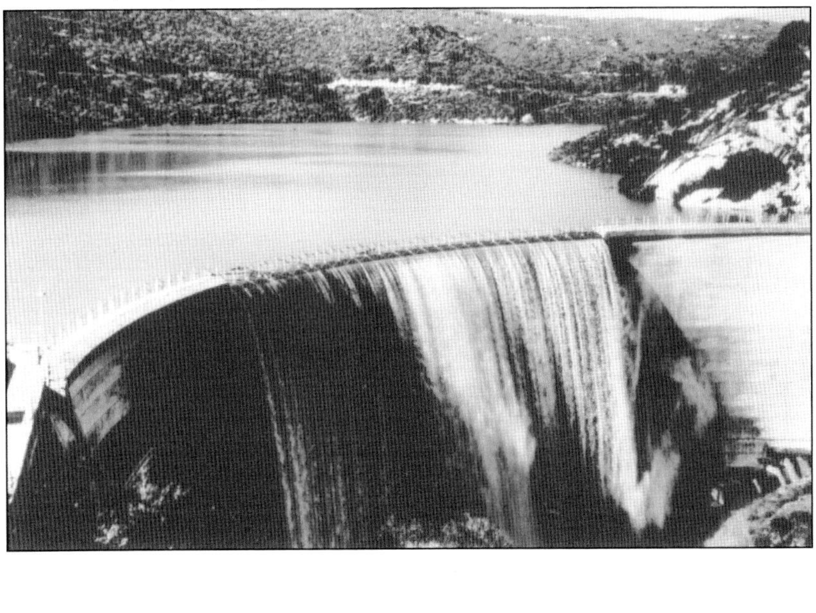

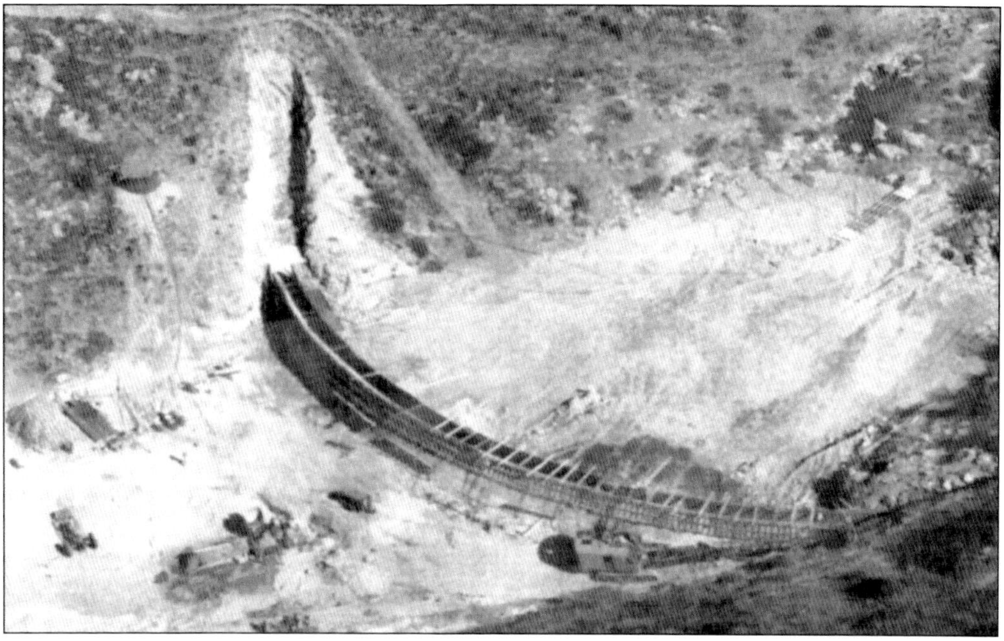

In 1946, George D. Clemson and John C. Wells agreed to combine their respective lands on the banks of San Marcos Creek to create a lake to conserve irrigation water for their farmlands. Wanting to keep the project in private hands, Clemson provided all the funding. They hired engineer John Covert to design and construct a dam that would conserve the area runoff that flowed into San Marcos Creek. Covert created San Marcos Dam, a concrete gravity arch structure that stood 45 feet high by approximately 270 feet long with a slim, four-foot-wide crest running along its length. The dam has no dedicated spillway; water simply flows over the thin crest. (Both, courtesy of San Marcos Historical Society.)

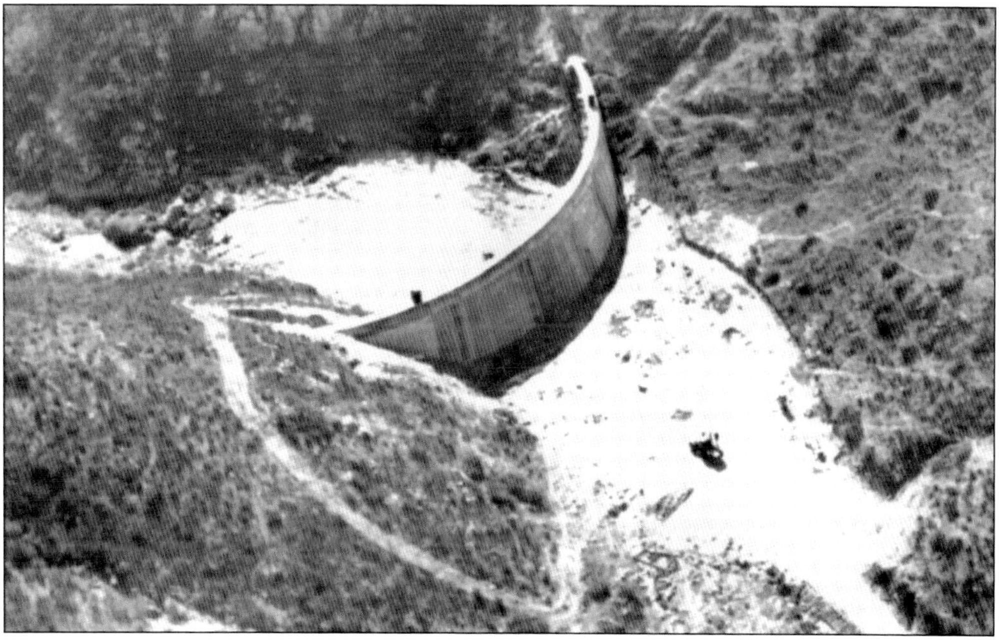

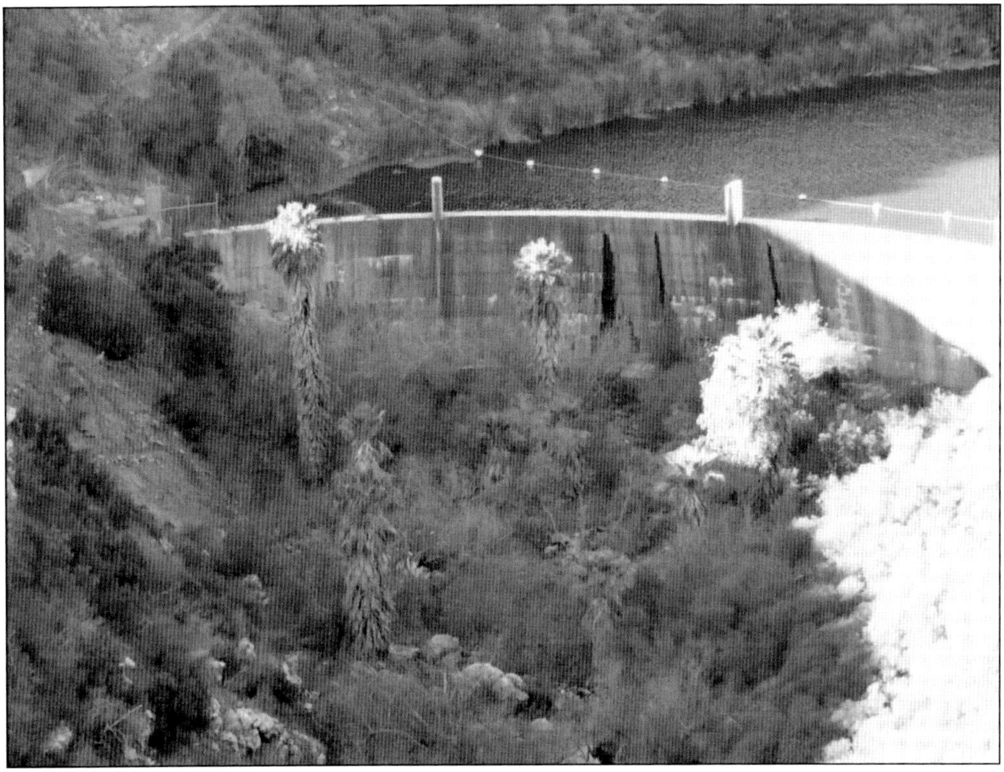

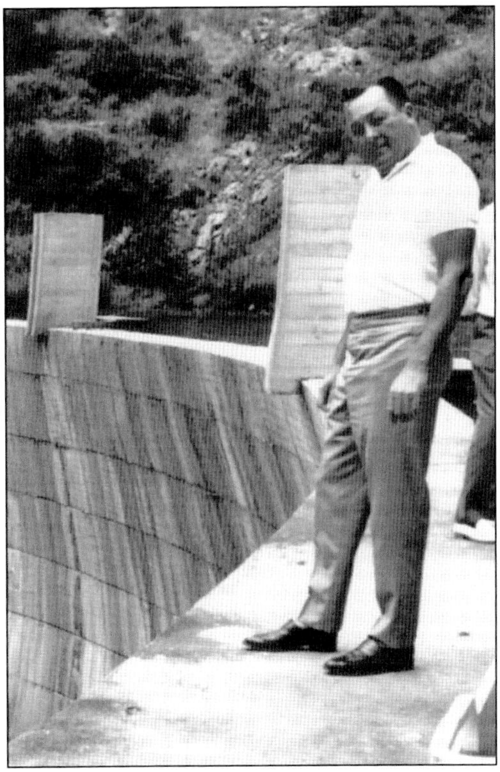

The narrow canyon site was ideal, but the unsettled streambed compelled Covert to seat the base of the structure on footers sunk 12 feet in the ground. San Marcos Dam is situated in a sage- and chaparral-covered gorge at the south end of the artificial lake it formed. The reservoir outlets into seasonally flowing San Marcos Creek, which flows west and releases into the Batiquitos Lagoon in Carlsbad. (Author's collection.)

In 1962, Gordon Frazar (pictured) and his brothers Bob and Don purchased 1,500 acres of the Clemson-Wells tract, including San Marcos Dam, to establish a lakeside community called Lake San Marcos. To enhance the development, they drained the old lake, increased its size, and reshaped it with indents to increase the number of lakeside properties. The shallow lake covers approximately 80 acres and is now a regional recreational facility. (Courtesy of San Marcos Historical Society.)

The Helix Water District constructed the Chet Harritt Dam across Quail Valley Creek in 1962 to create Lake Jennings. Jennings would be a storage facility for water received from the first San Diego Aqueduct. The 9,800-acre-foot lake is the centerpiece in delivering water through the nearby treatment plant to over 260,000 consumers in the communities of Lemon Grove, La Mesa, and Spring Valley. (Author's collection.)

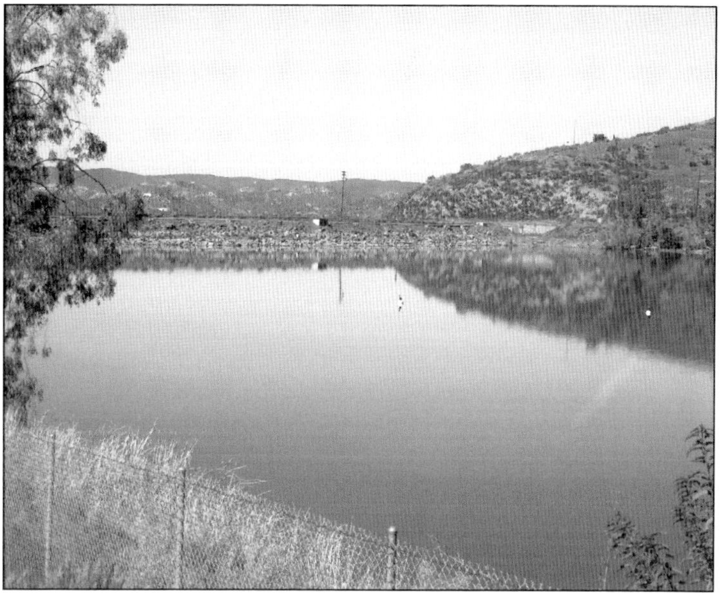

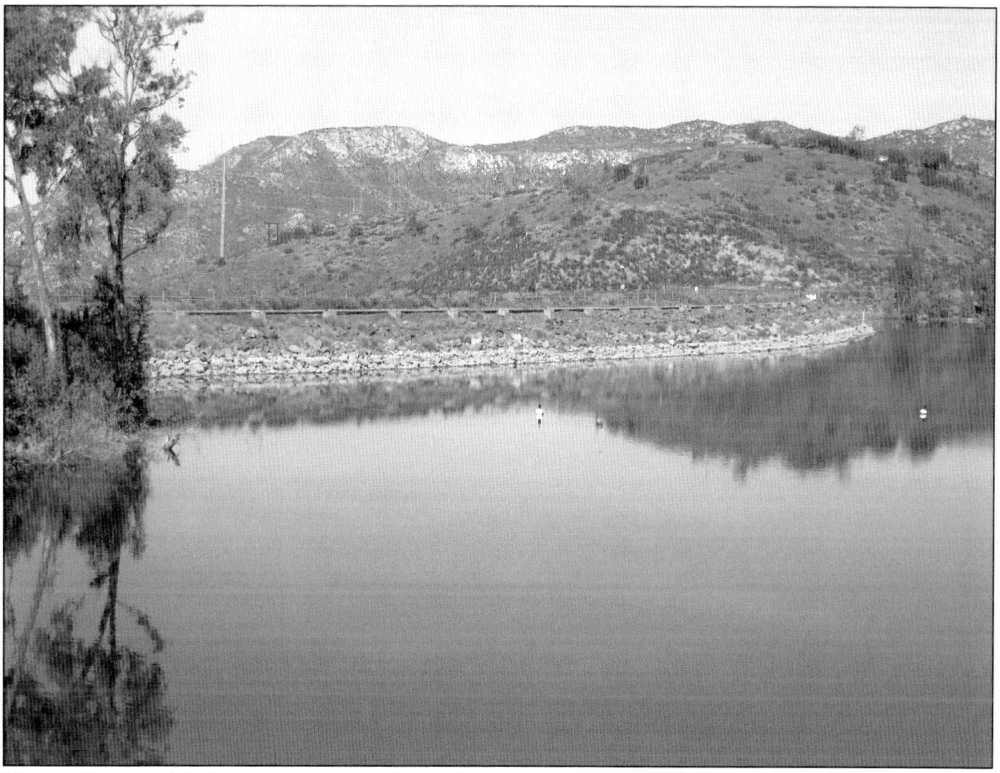

Harritt Dam and Lake Jennings sit on lands that were within Rancho El Cajon during California's Mexican era. Helix Water District officials named the lake for Bill Jennings, who served on the California Water Commission, was the chief legal counsel for the district, and was the water district's first manager. The dam was named after longtime employee Chester Harritt. (Author's collection.)

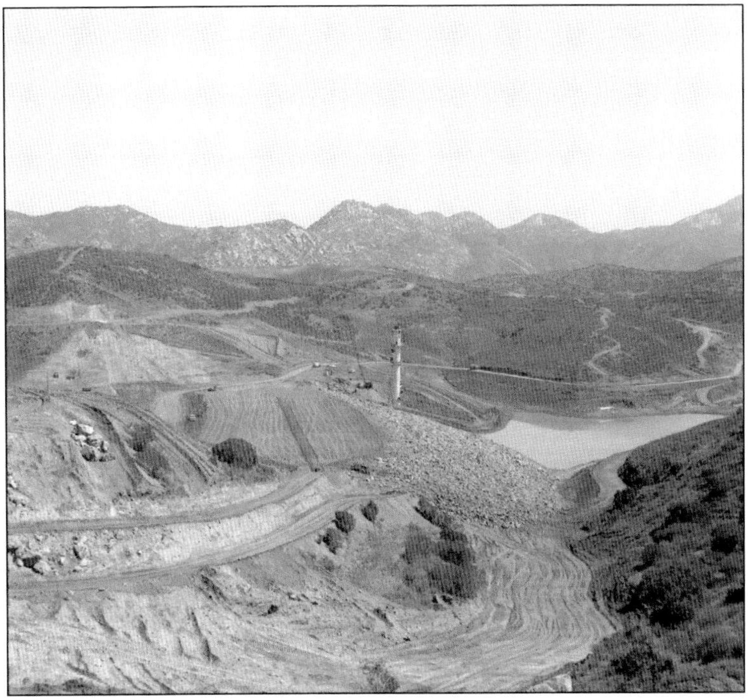

This construction image from 1962 show the rock course laid on the upstream side as heavy equipment compacts the earth fill material to form Harritt Dam. Engineers shaped the reservoir around the outlet tower. Lake Jennings Road transits the downstream face of the 200-foot-high earthen dam. With no significant natural runoff, the district designed the reservoir as a storage facility. (Courtesy of Helix Water District.)

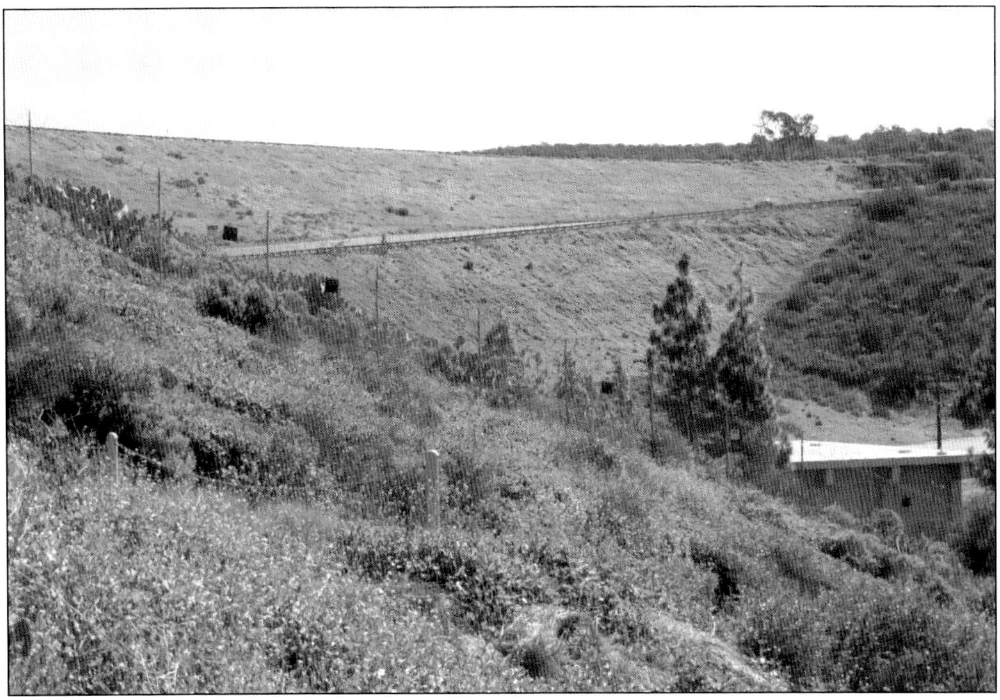

The downstream face of Harritt sprouts grass and weeds in the wet months and dries up in the summer. A two-lane paved road leading to the Lake Jennings Campground crosses the dam's crest. Manufactured homes now surround the artificial lake, which is home to 195 species of birds, including turkey vultures, aquatic birds, and white pelicans. A state proposal appropriating $40 million for the reconstruction of the dam is pending. (Author's collection.)

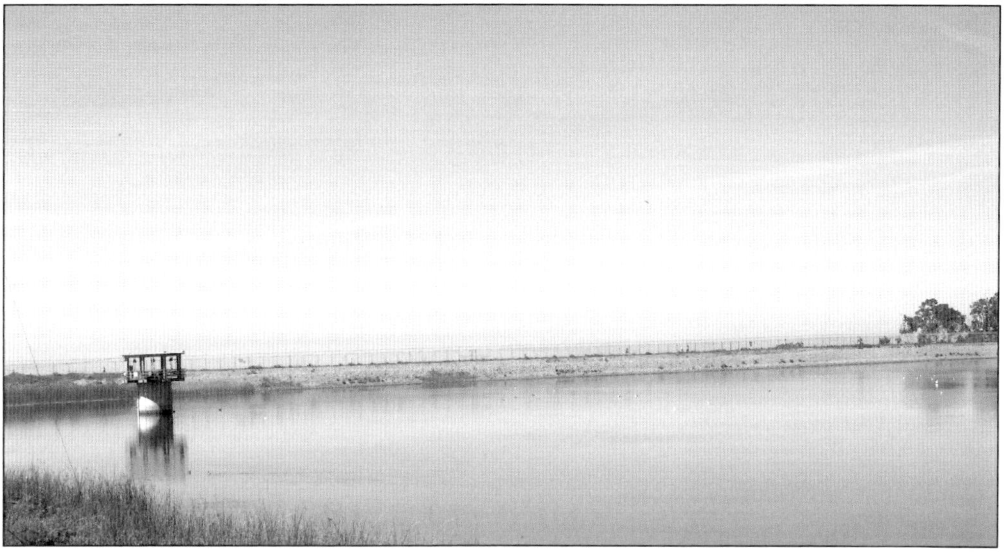

Miramar Dam and Reservoir are located in the community of Scripps Ranch on the north edge of San Diego. The city completed the dam in 1960 as part of the second San Diego Aqueduct project. With only a one-square-mile catchment, Miramar is a storage reservoir that can accommodate 6,682 acre-feet of water from the Colorado River and the California Aqueduct. The city installed a water treatment plant immediately below the reservoir in 1962 that can treat 140 million gallons per day and performed other upgrades in 2011. Today, the reservoir and adjacent park are popular recreation sites. The dam's paved 12-foot-wide crest is open during the day to the public for jogging, walking, and biking and enjoying the spectacular view to the coast. (Both, SDCWHA.)

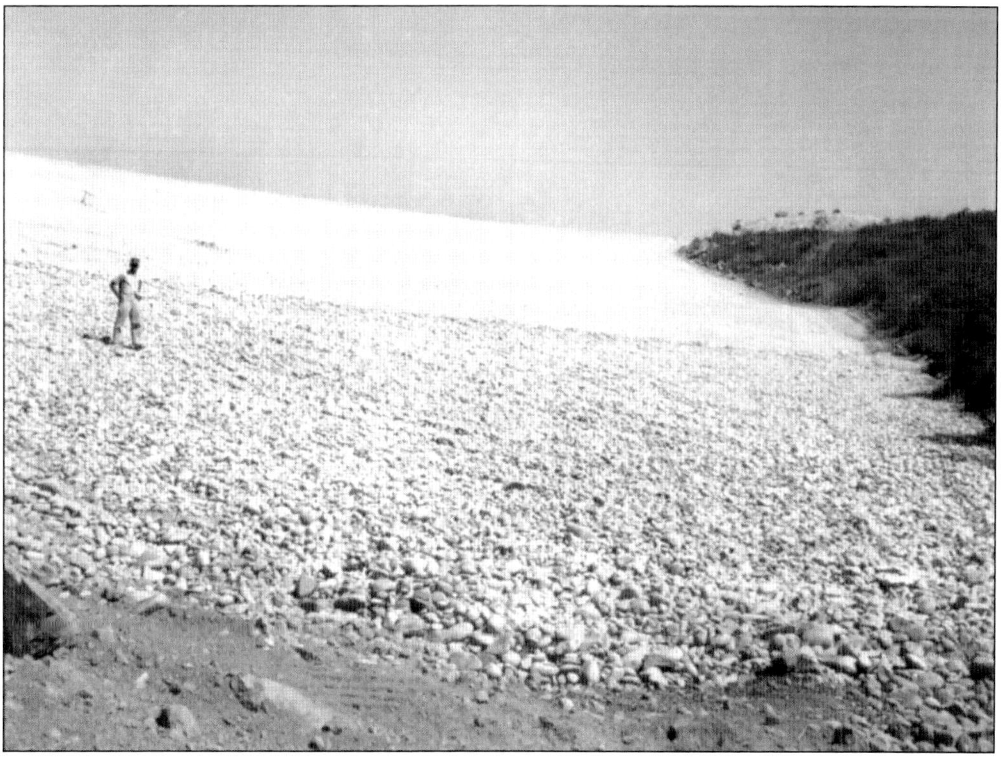

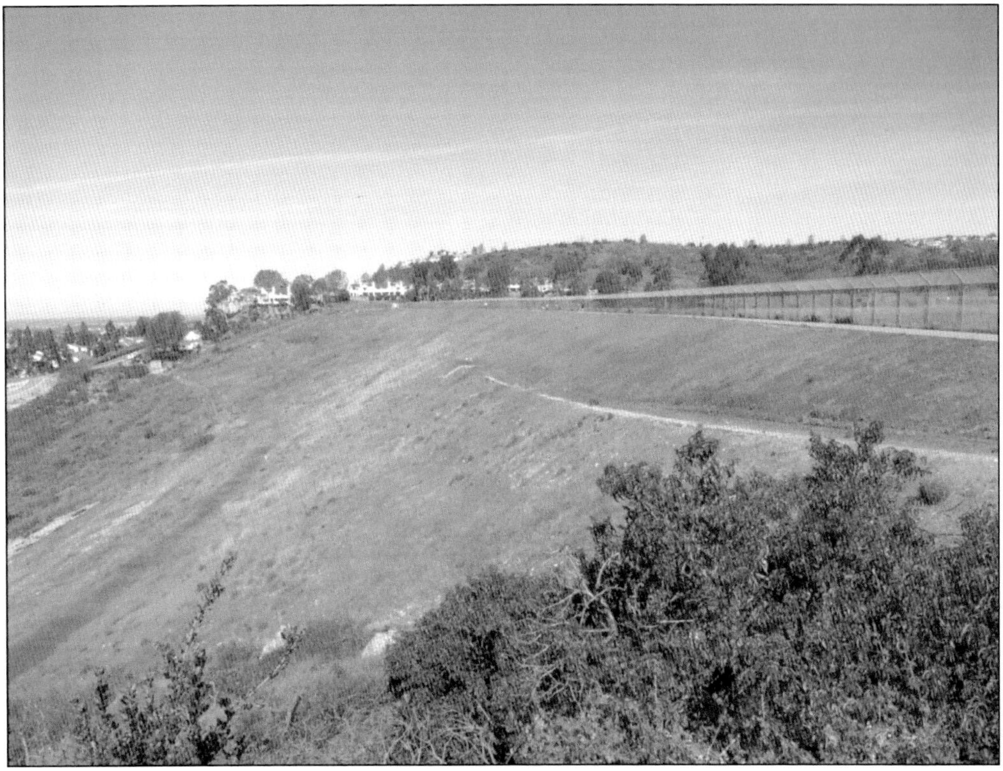

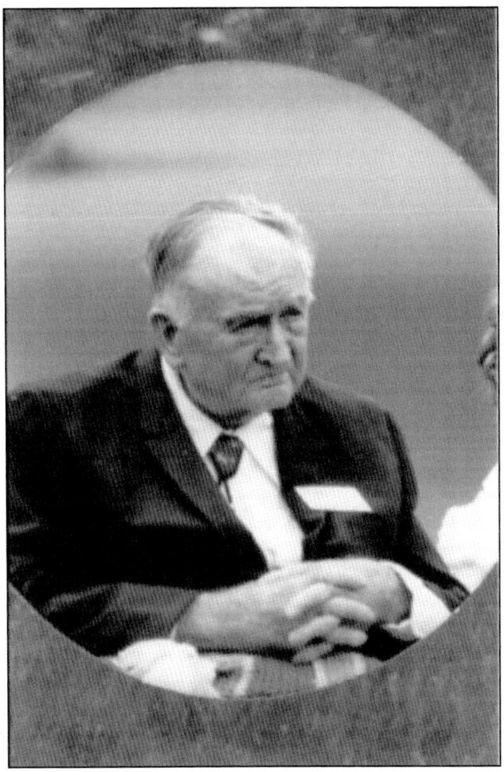

Newspaper publisher E.W. Scripps purchased 400 acres of barren, sage-covered mesa some five miles from the coast in the 1890s for $5,000. Scripps named the area and his ranch Miramar, which means "sea view," for the unobstructed view to the ocean. The family constructed seven dams on the property for irrigation water. One, called Surr Dam, was below the site of the present Miramar Dam. (Author's collection.)

Dixon Dam takes its name from the original property owner and longtime water district employee James Dixon. Dixon arrived with his family in the Escondido Valley in 1888. As a teenager, he found employment with the Escondido Mutual Water Company as a "ditch man," eventually became a supervisor, and remained with the company for over 40 years. (Courtesy of Escondido Library, Pioneer Room.)

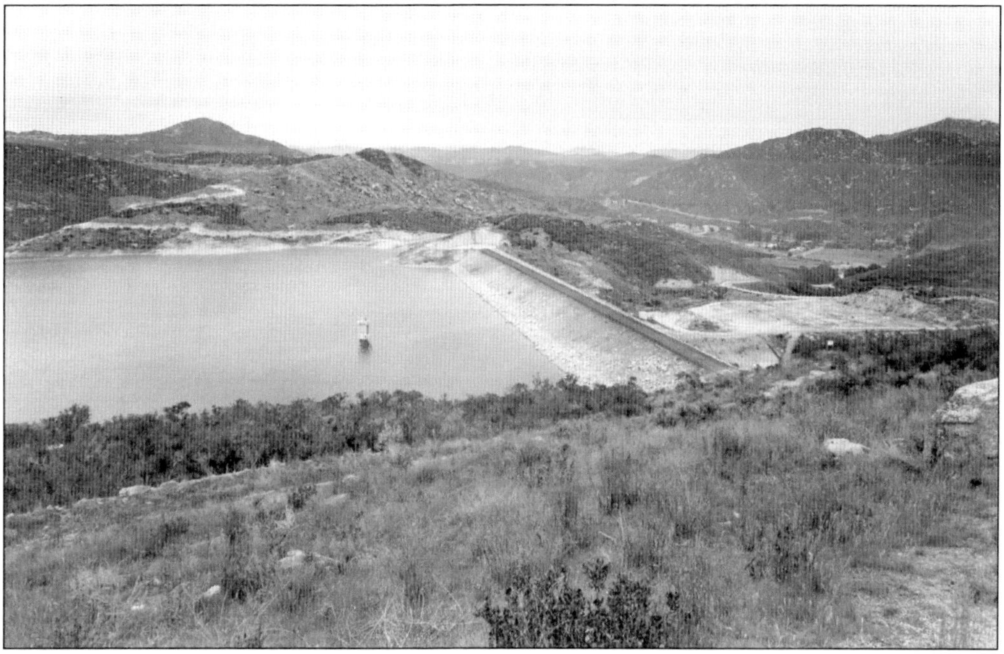

In 1910, Dixon constructed a small earth dam near his ranch to provide drinking and irrigation water. In 1955, the Escondido Water District officials made an offer to Dixon for that dam and the surrounding 250 acres of land, but they did not acquire the property until the early 1960s, for $25,000. The Oberg Construction Company built the city-designed, 100-foot-high, 910-foot-long structure northeast of Escondido in 1971 at a cost of $1.7 million. Dixon was a terminal storage basin for domestic water from the San Diego Aqueduct and for flood control and recreation. In 1977, the city added a filtration plant to the system. The reservoir is renowned as a fishing destination with record-sized largemouth bass. (Both, courtesy of Escondido Library, Pioneer Room.)

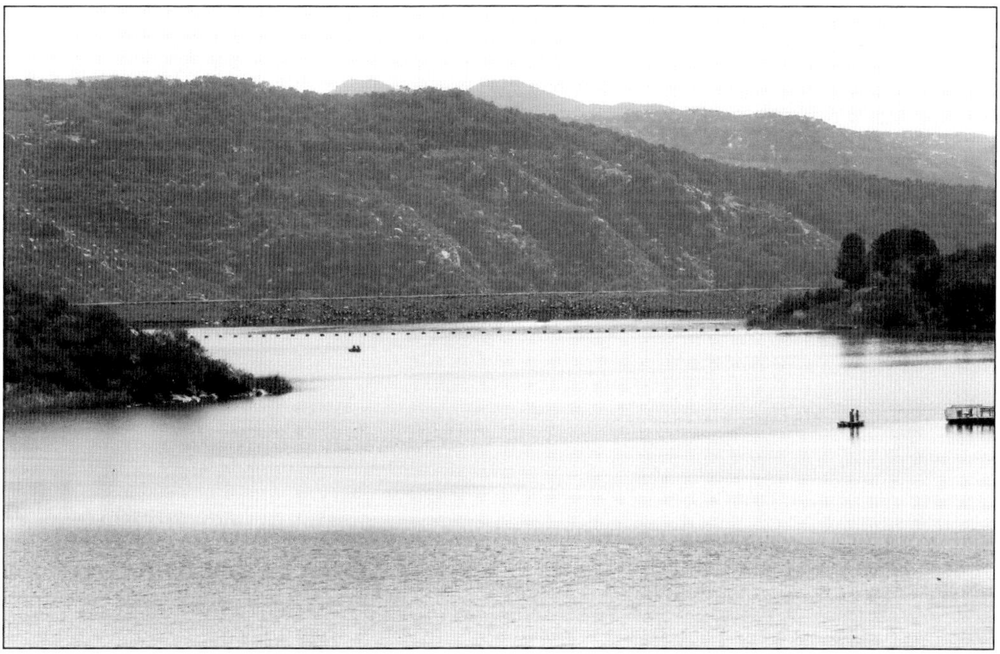

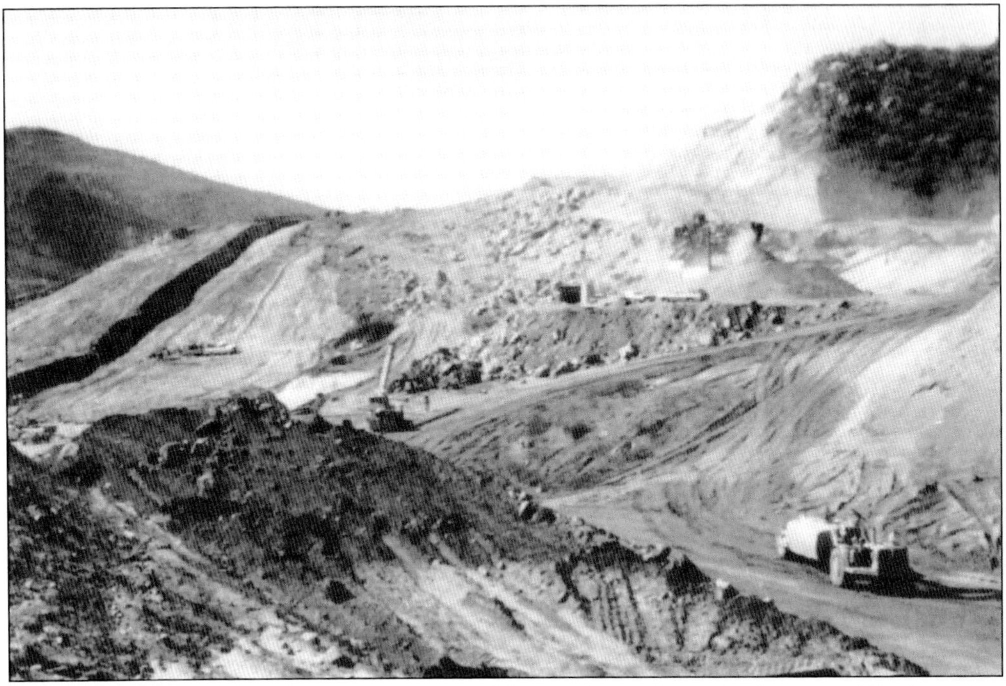

The Valley Center Water District constructed Turner Dam through a Bureau of Reclamation agriculture loan. The district named the dam after longtime board member Robert J. Turner. The rock-fill dam has a distinctive oversized concrete channel spillway, and a district service road cuts the downstream face of the dam and transits the 40-foot wide crest. (Courtesy of Valley Center Water District.)

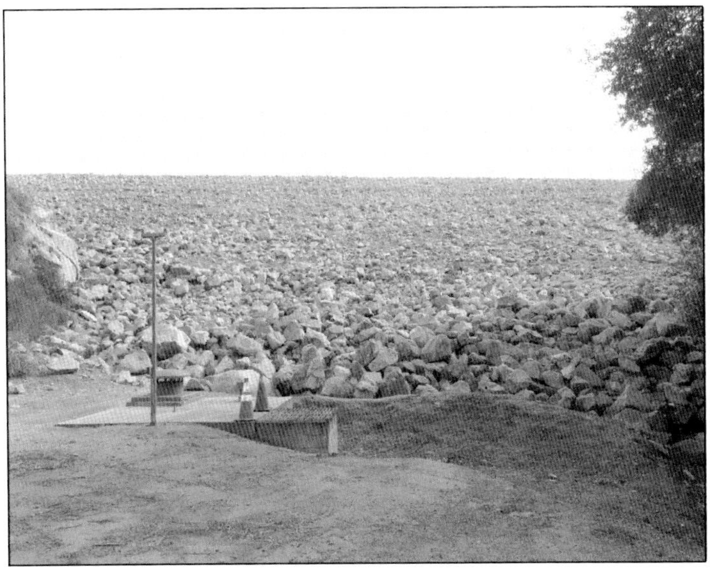

The dam, which conserves Moosa Creek, was originally designed to store untreated Colorado River water and be an emergency storage facility. But by the early 1980s, the unrestrained growth of organic matter rendered the reservoir water undrinkable and unsuitable for agricultural use. Since the algae outbreak, the 1,612-acre-foot-capacity reservoir has no water consumptive purpose or recreation use. (Author's collection.)

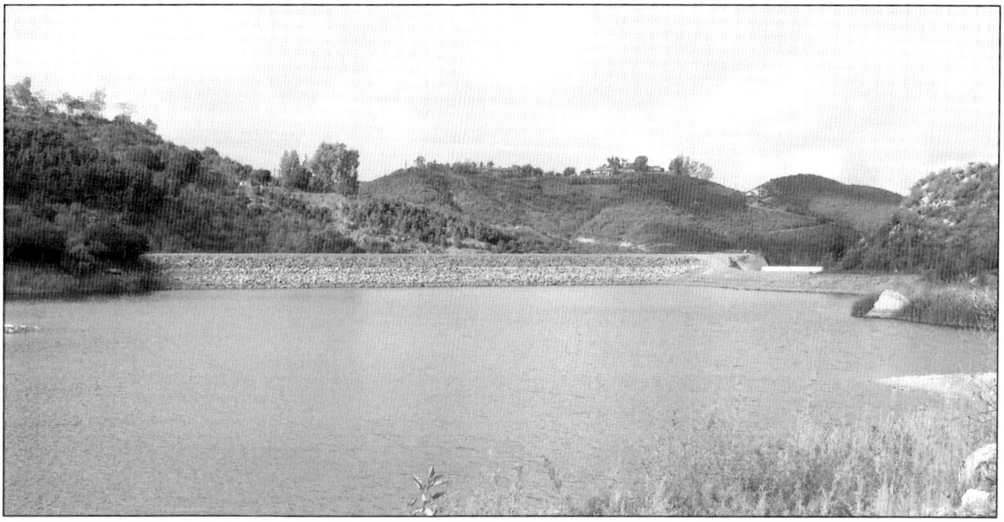

The fenced reservoir is closed to public use and guarded to prevent trespassers from poaching the trophy bass and catfish. The reservoir still exists because of the fish population, the rich avian presence, and the expense of removing the dam. To keep the lake healthy for fish and wildlife, the water district placed two solar-powered aeration units in the lake to continuously agitate and circulate the waters. (Author's collection.)

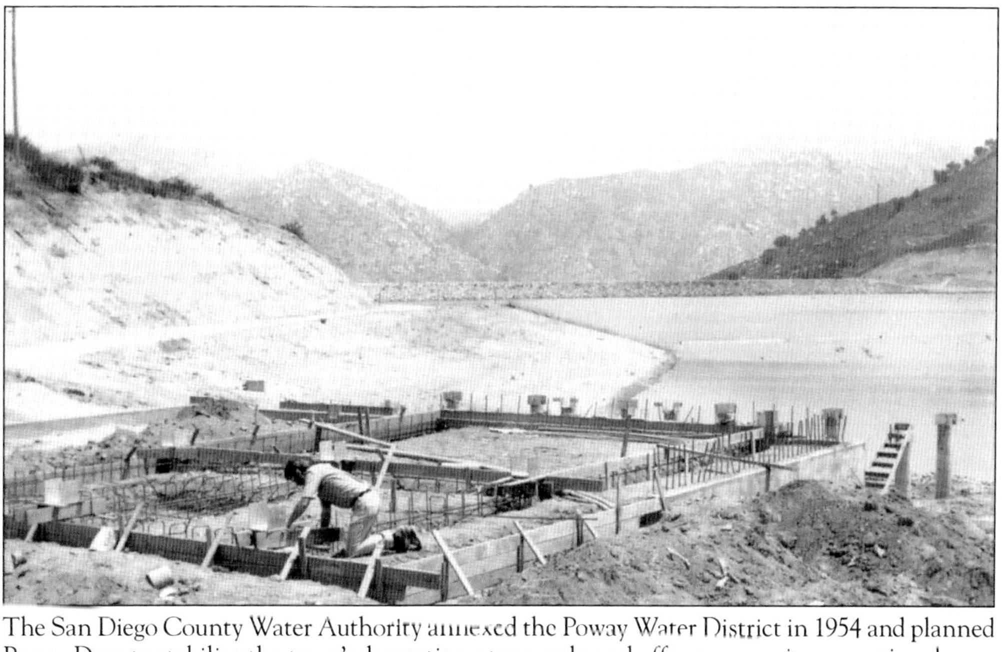

The San Diego County Water Authority annexed the Poway Water District in 1954 and planned Poway Dam to stabilize the town's domestic water supply and offer community recreational uses. A construction bond for the dam failed in 1964 and again in 1966, but in June 1969, it received 87 percent of the vote to go forward. Work on the dam started in December 1970 and concluded in 1972. (Courtesy of Poway Historical Society Museum.)

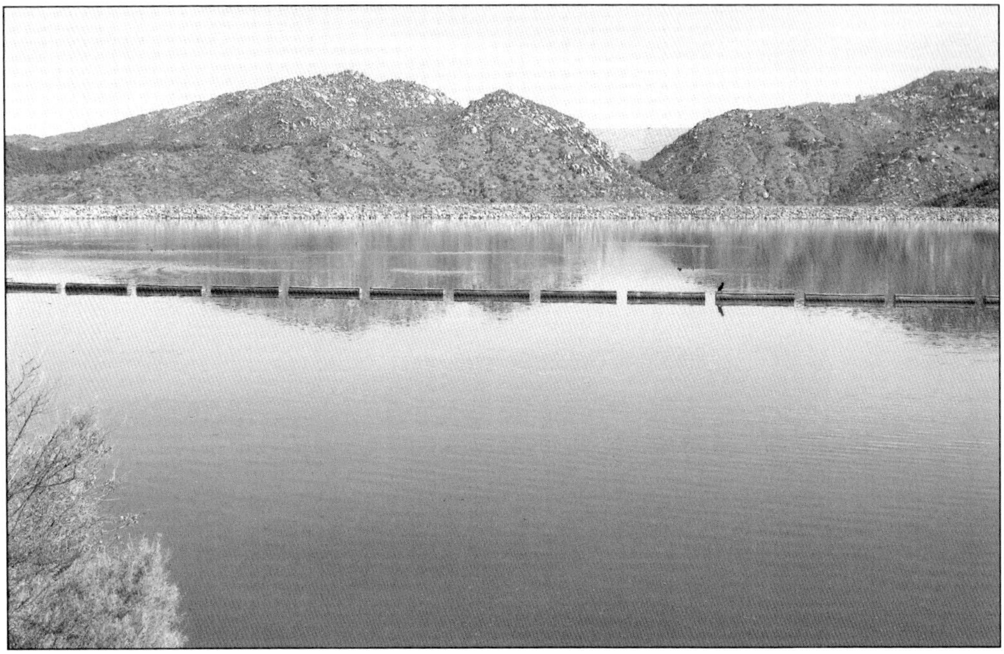

The earth-and-rock Poway Dam is 162 feet high and 970 feet long, with a service road atop the 20-foot-wide crest. The dam has no formal spillway. The 3,330-acre-foot-capacity reservoir receives its water from the San Diego County Water Authority and transfers it to the nearby Lester Berglaund Water Treatment Plant. Ramona Dam is visible, three miles distant in the background in the notch between the hills. (Author's collection.)

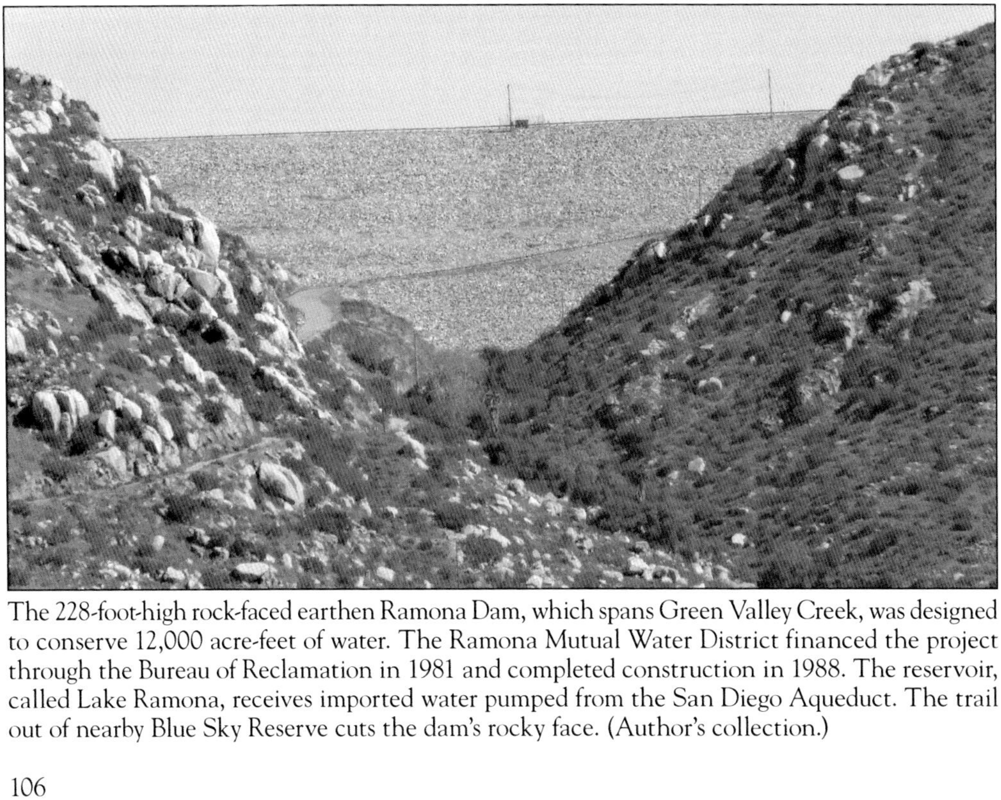

The 228-foot-high rock-faced earthen Ramona Dam, which spans Green Valley Creek, was designed to conserve 12,000 acre-feet of water. The Ramona Mutual Water District financed the project through the Bureau of Reclamation in 1981 and completed construction in 1988. The reservoir, called Lake Ramona, receives imported water pumped from the San Diego Aqueduct. The trail out of nearby Blue Sky Reserve cuts the dam's rocky face. (Author's collection.)

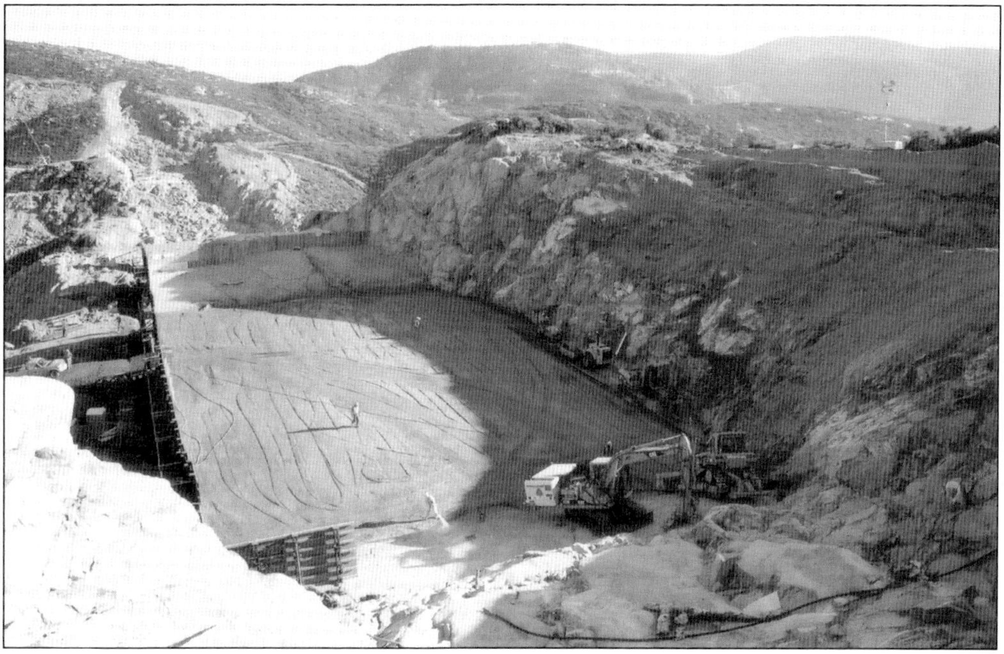

Olivenhain Dam was the first gravity dam in California constructed with rolled-compacted concrete. Kiewit Pacific Company started work on the dam in June 2001 and completed the $200-million project in July 2013. The dam stands 318 feet in height, runs 2,400 feet in length, and sits within the 700-acre Elfin Forest Recreational Reserve. It is part of the county's Emergency Storage Project. (Courtesy of San Diego County Water Authority.)

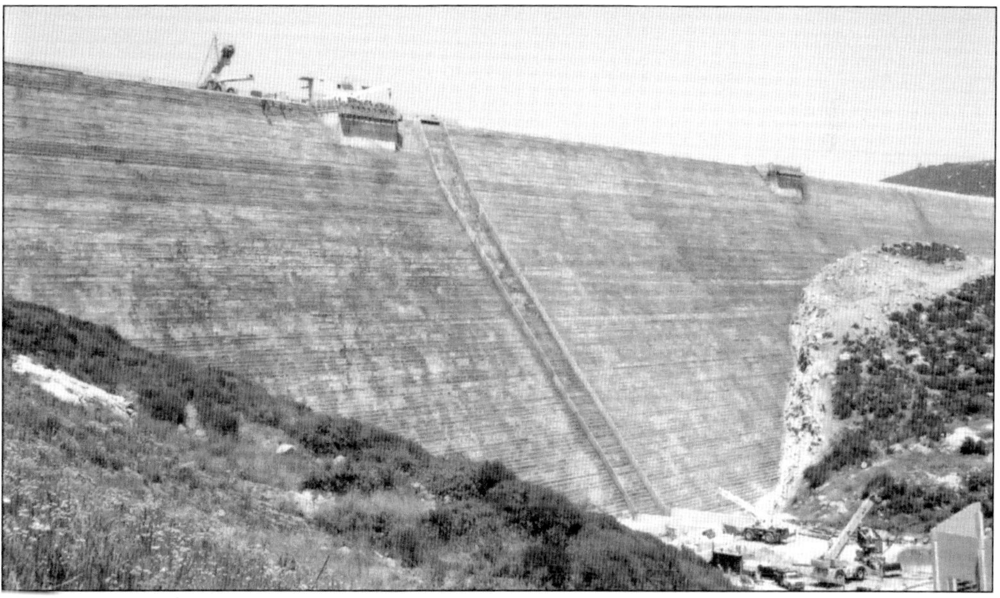

Granite quarried during the excavation for the dam's foundation, which took a year, was the primary construction material. The dam's face is stained with a desert varnish shade to blend with the surroundings. Because Olivenhain is not situated on a waterway, it receives water piped from nearby Lake Hodges and from the second San Diego Aqueduct. (Courtesy of San Diego County Water Authority.)

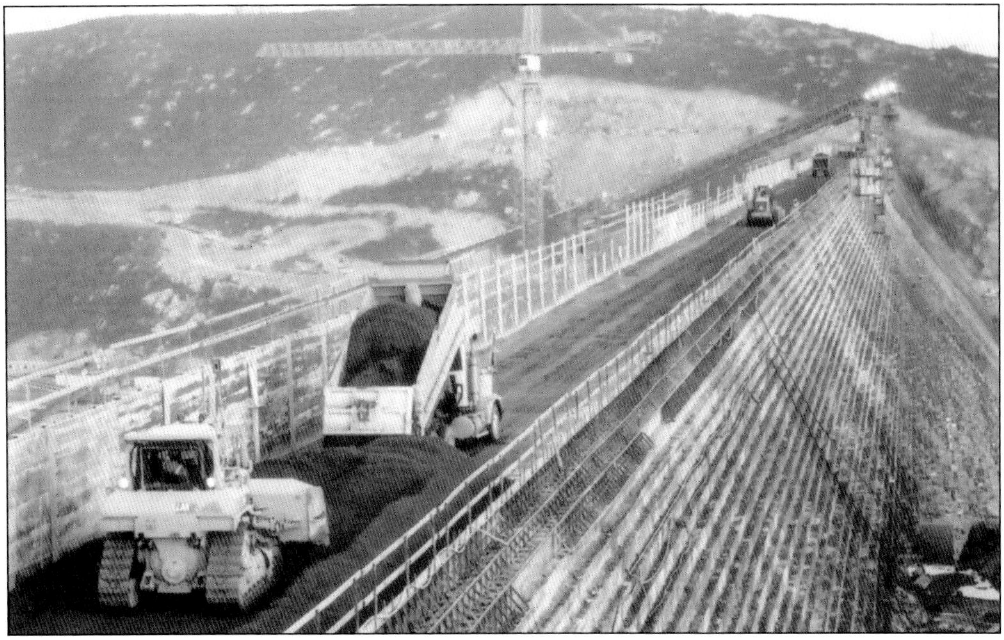

Olivenhain Dam required 1.4 million cubic yards of compacted concrete to be placed in layers across the structure. State testing proved the rolled-compacted method to be both more economical than a conventional concrete dam and quicker to build. This process required crews to work 24-hour days six days a week to ensure proper bonding of the concrete, which workers placed in 300 twelve-foot-thick layers. (Courtesy of San Diego County Water Authority.)

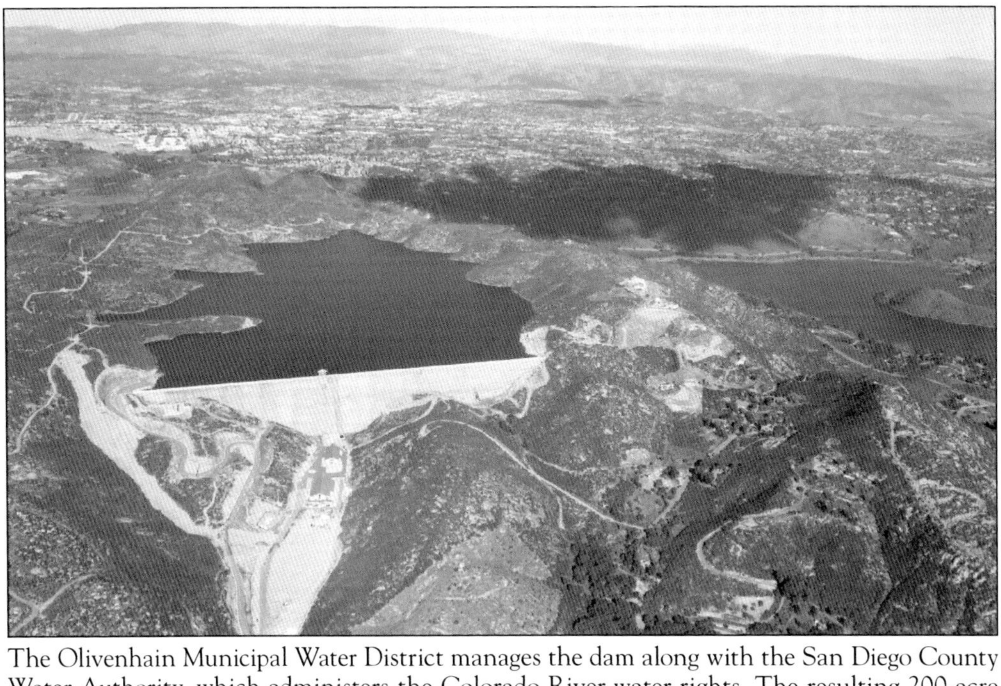

The Olivenhain Municipal Water District manages the dam along with the San Diego County Water Authority, which administers the Colorado River water rights. The resulting 200-acre reservoir can conserve 24,000 acre-feet of water, enough to supply 50,000 families of four with water for a year. Lake Hodges is on the right. (Courtesy of San Diego County Water Authority.)

Two

The Dams That Never Happened

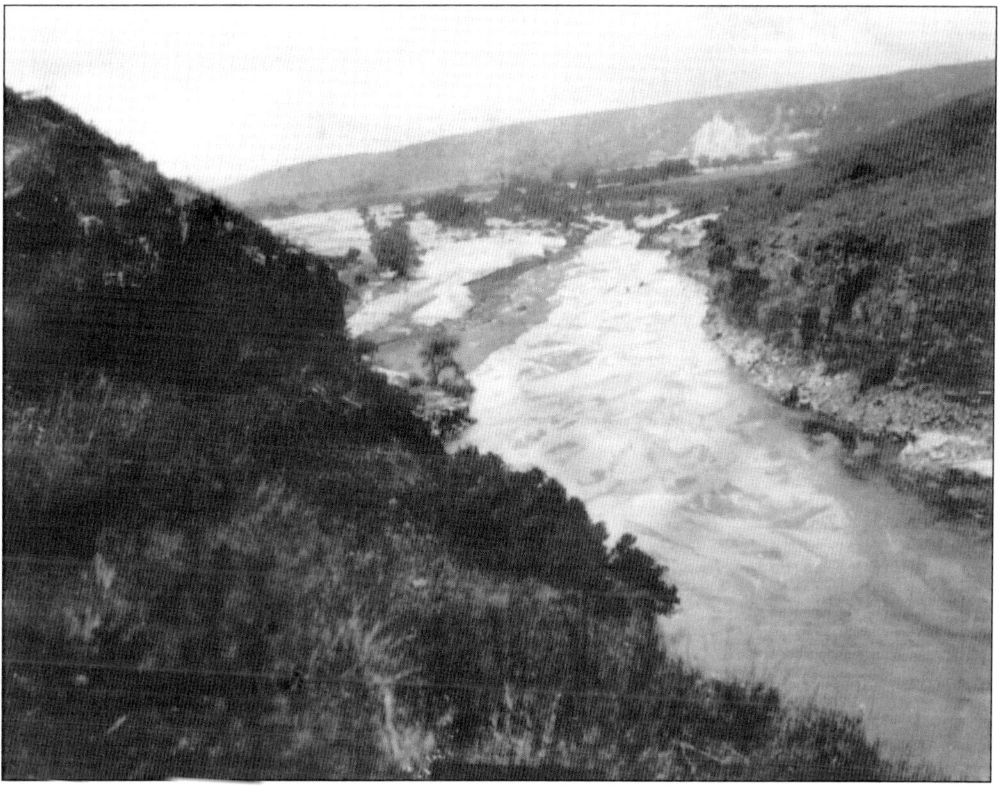

In the first decades of the 20th century, private water entrepreneurs and municipal water officials scoured the county's river systems seeking dam sites to create storage reservoirs. Some of these enterprising men hoped to profit, and some hoped to serve their communities. Some projects went forward, while others did not. This 1916 view shows a potential but unrealized site on the San Diego River where it flows through Mission Gorge. (SDCWHA.)

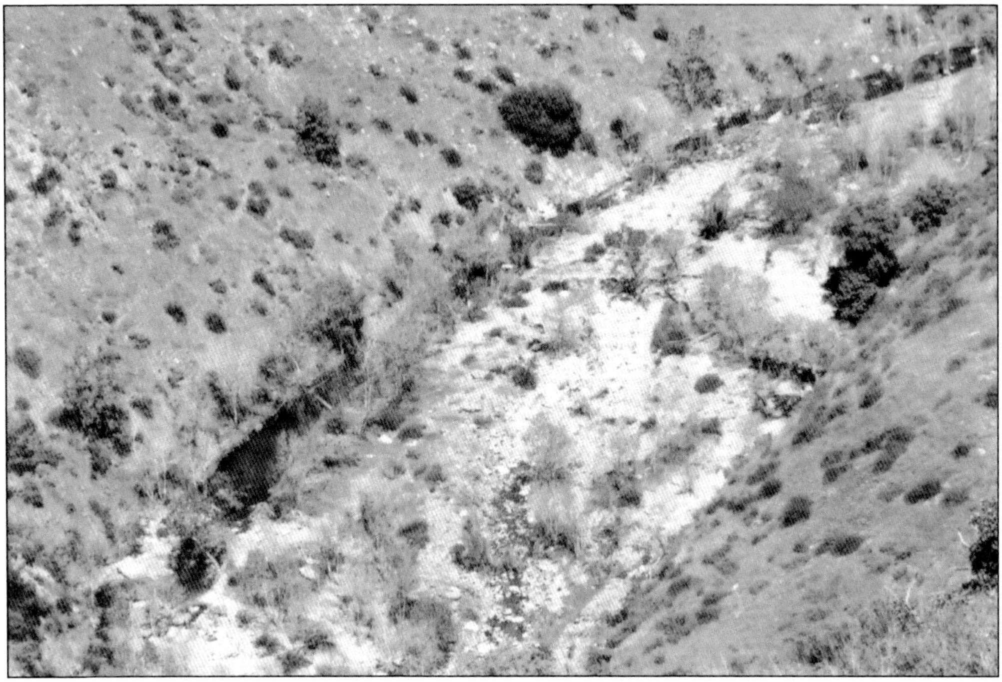

For 30 years, Ed Fletcher pursued his namesake dam. To maximize the efficiency of his Cuyamaca Water Company, Fletcher considered constructing as many as eight dams and associated reservoirs along the San Diego River, with Fletcher Dam the keystone structure. Fletcher envisioned his dam about 3,600 feet upstream from the existing diverting dam in a narrow canyon called Rocky Bar. At 840 feet of elevation, his engineers assured him the water collected behind the dam could pass by gravity flow to any part of the city. Thomas King, an engineer for the Cuyamaca Water Company, drew up plans for a 160-foot-high multiple-arch dam that would cost $762,900. The reservoir, Fletcher Lake, could store a million gallons of water that could be fed into the flume system when needed. The completion of El Capitan in 1934 rendered Fletcher's dream moot. (Both, SDCWHA.)

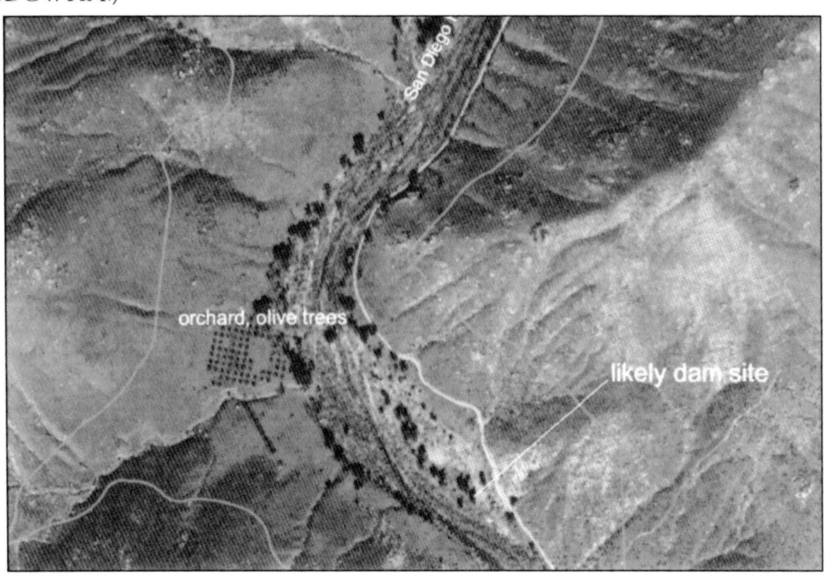

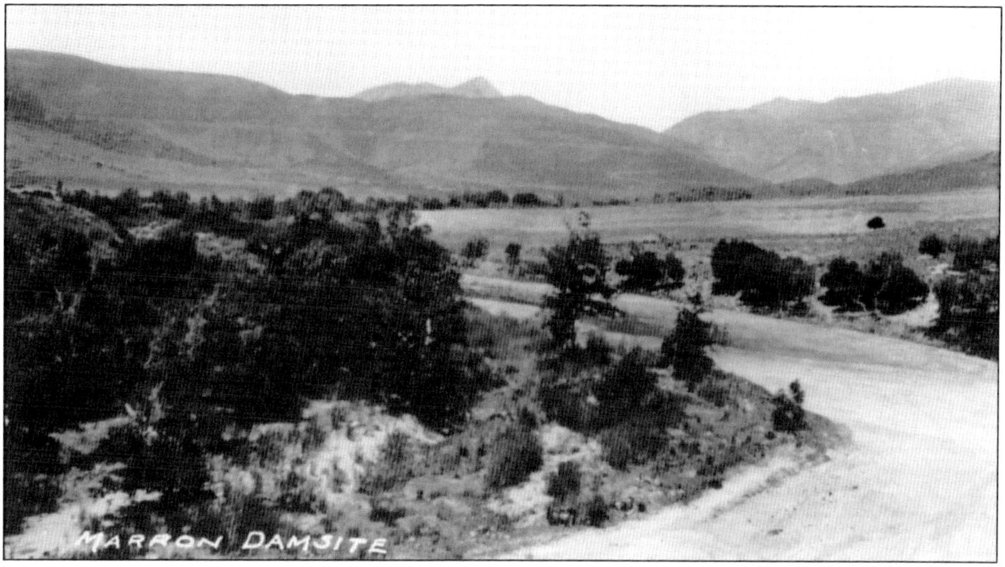

The Marron Dam project was the most grandiose of all the dam schemes that never happened. In the 1900s, city hydraulic engineer H.A. Whitney discovered potential dam sites in the Marron Valley and Pine Valley in the southern part of the county, close to existing reservoirs. Whitney favored the Marron Valley site because it could be connected via an aqueduct to the city-owned Lower Otay Reservoir, and the city already owned the property and the water rights to the adjacent creeks. Whitney speculated that a 110-foot dam constructed at the confluence of the Tijuana River and Cottonwood Creek could potentially conserve 12 billion gallons of water. Unfortunately, the proposed dam and reservoir straddled the US-Mexican border, and the issues of funding and distribution ratios proved unsolvable. The barbed wire fence stretching across the river in the image below marks the border. (Both, SDCL.)

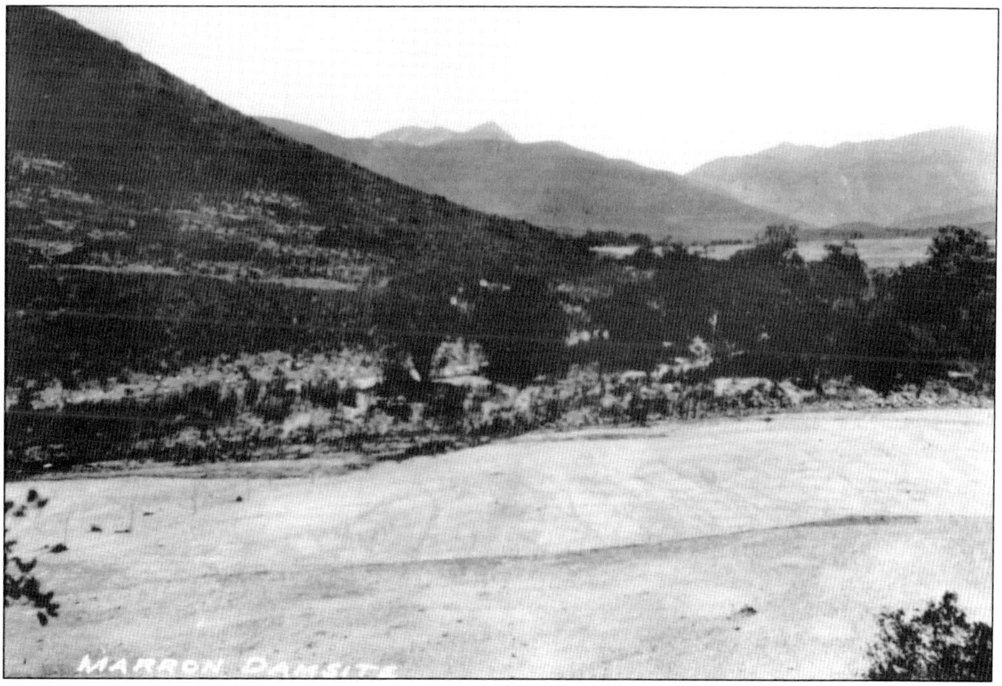

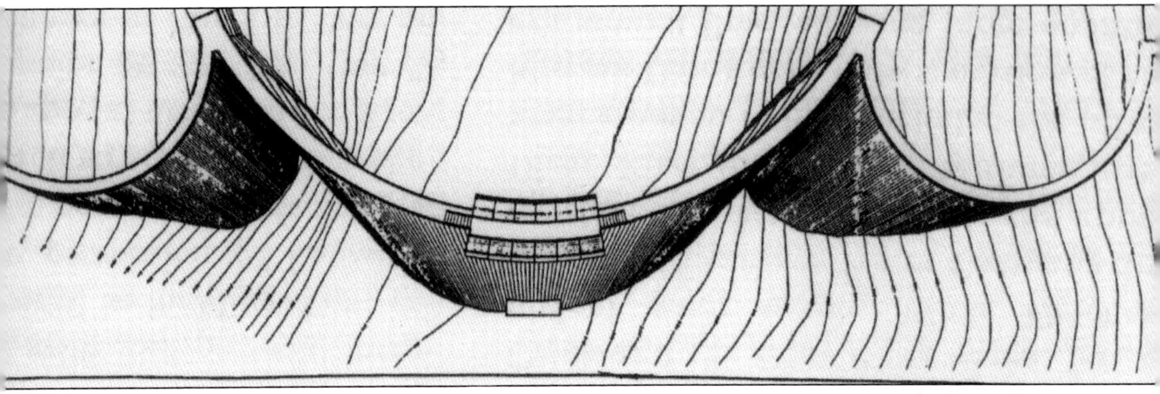

According to Ed Fletcher, the funding for a San Elijo Dam would be a shared scheme between the Santa Fe Railway, the San Dieguito Mutual Water Company, and the Cardiff Irrigation District. John Eastwood actually drew designs that called for a 140-foot-high "radial cone" dam with three arches. It would span Escondido Creek where it enters the San Elijo Lagoon, just south of Encinitas and about a half-mile from the coast. W.L. Huber filed for the land in March 1917 in the name of Fletcher and the San Dieguito Mutual Water Company. Fletcher's plan was to create a reservoir that could conserve 12,000 acre-feet of water, which he could then sell to coastal consumers. Although the state Department of Engineering approved the project, four local landowners actively protested the filing. Fletcher abandoned San Elijo when he failed to organize the consortium and raise the $30,000 needed for construction. (SDCWHA.)

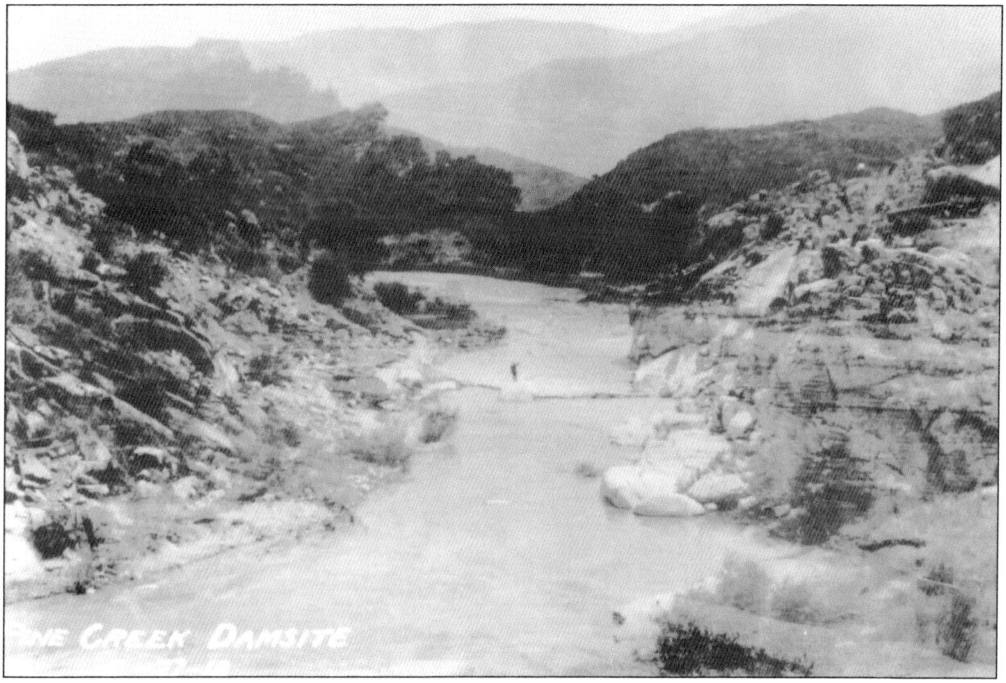

As early as 1910, city water experts considered building a dam in Pine Valley on the upper Cottonwood catchment. The site they preferred was on Pine Creek near the confluence of Cottonwood Creek. However, the site's rugged and remote location and small catchment delayed the decision until the construction of Barrett Dam ended the project. This photograph shows the proposed dam site in 1918. (SDCL.)

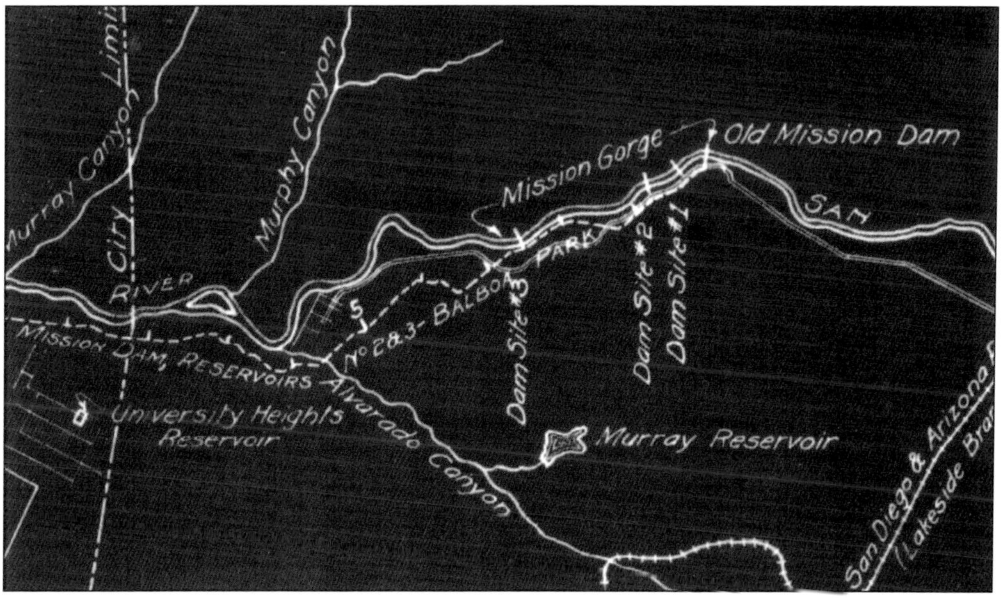

Of all the failed projects, the construction of a dam in Mission Gorge seemed most likely to succeed. There were three sites; of the two likely sites, Fletcher owned one and the city the other. Both gorge dam sites were conveniently located only miles outside the city. After two decades of debate, the decision to build El Capitan squelched these projects. (SDCL.)

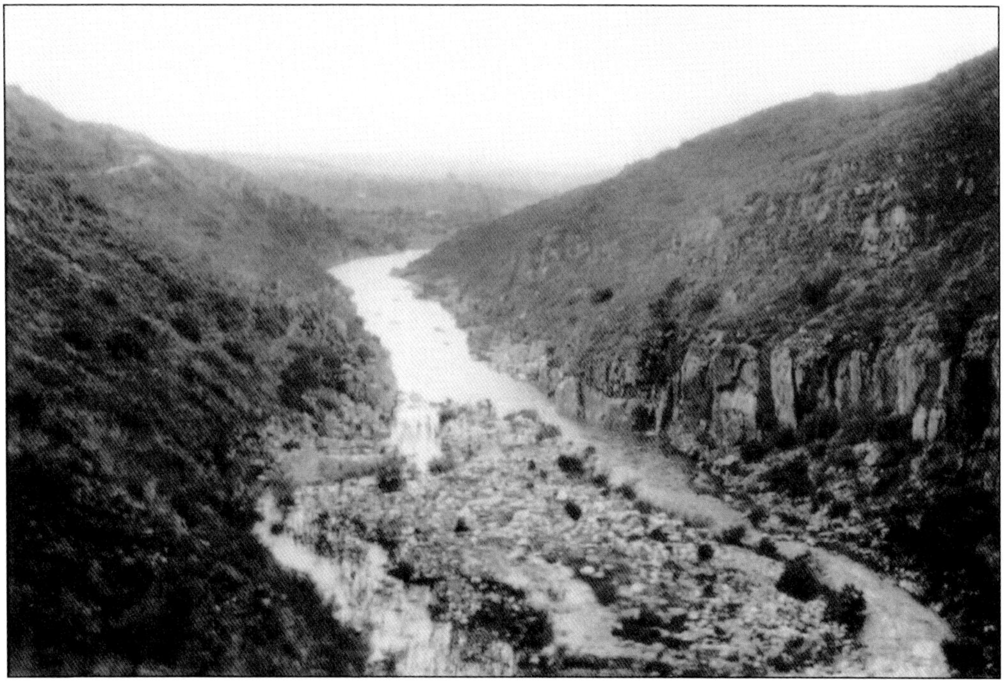

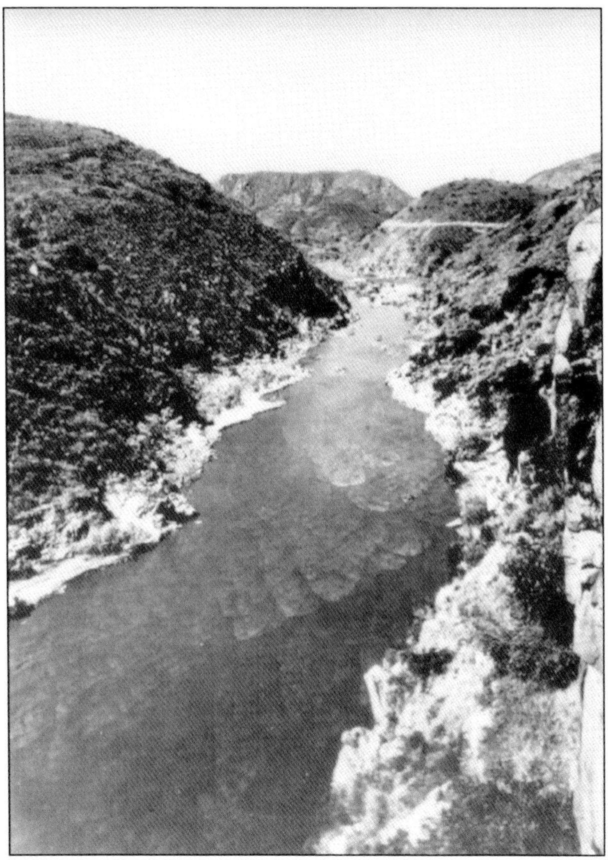

One gorge site was about a half-mile above the mouth of the gorge at the western end. The site featured exposed rock in the streambed and walls of the narrow canyon. Engineer C.W. Earl called it "the most ideal dam site in this section of the state." Multiple-arch dam expert John Eastwood prepared plans for a radial cone dam with an estimated construction cost of $550,000. (SDCL.)

This 1921 drawing shows the second gorge site, approximately half a mile below the old Mission Dam. Another engineer called this the most logical site for a dam. Lars Jorgenson drew plans for a 230-foot-high dam with three arches. But a dam here would have had evaporation issues and submerged the town of Santee and 5,000 acres of prime agricultural land. (SDCL.)

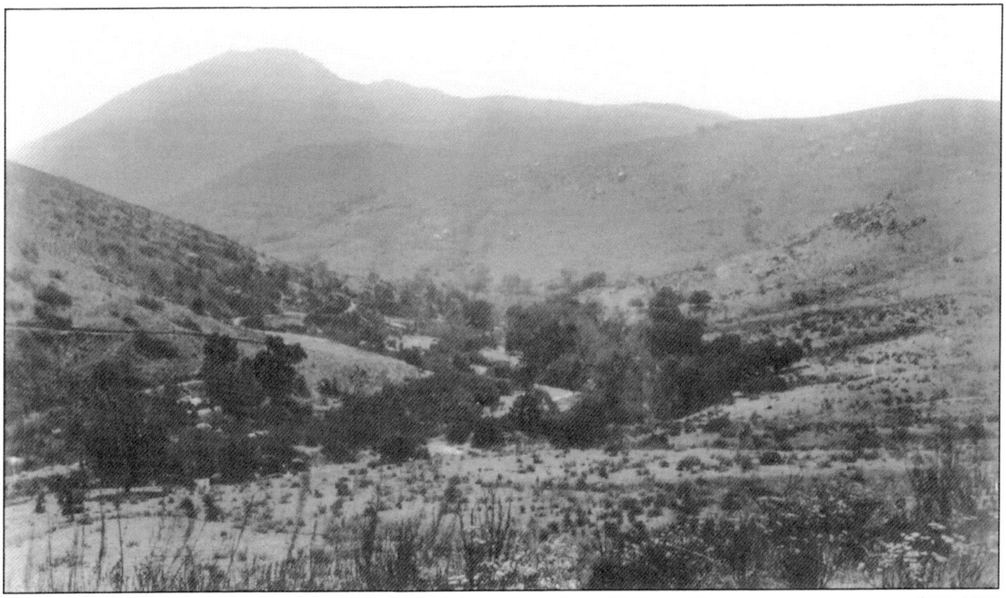

In the early 1900s, Fletcher, through his Volcan Land and Water Company, spent over $30,000 coring and drilling to determine if he could build a sizable concrete dam in Pamo Valley. Situated about seven miles north of the town of Ramona, water came from Santa Ysabel and Temescal Creeks, which bookended the valley between Black Mountain, Pine Mountain, and the Orosco Ridge. (Courtesy of UCSD Special Collections.)

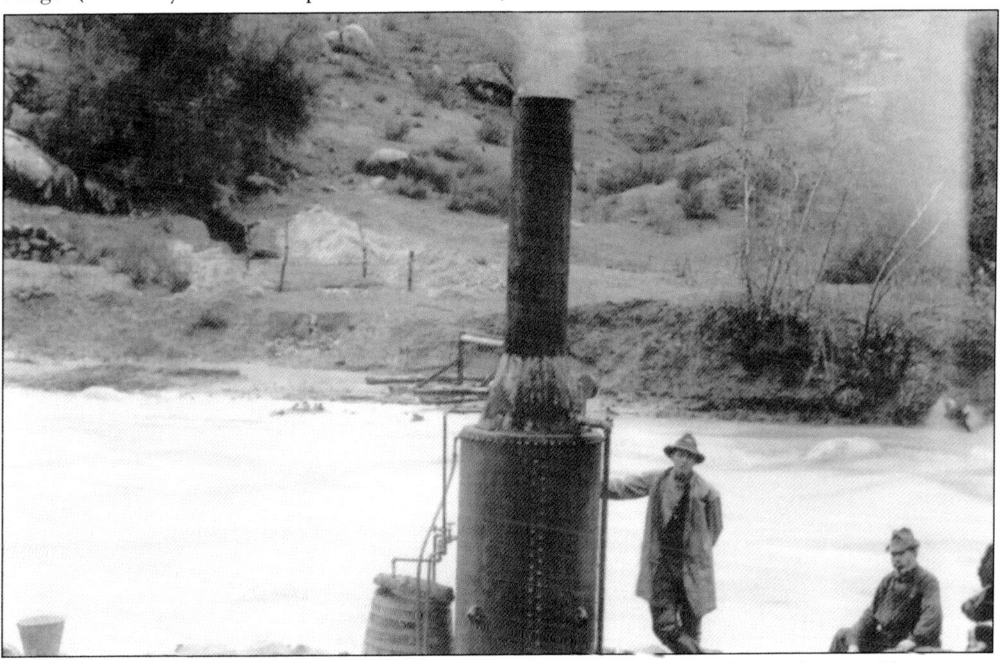

When Fletcher's Pamo venture waned, the City of San Diego stepped in and started buying up portions of the valley in 1925. City planners considered first a 156-foot-high dam, then a 264 foot concrete dam. Either reservoir would have submerged the pristine valley, but the arrival of Colorado River water in the 1940s ended the scheme. The city now leases the valley as pasturelands to local ranchers. (Courtesy of UCSD Special Collections.)

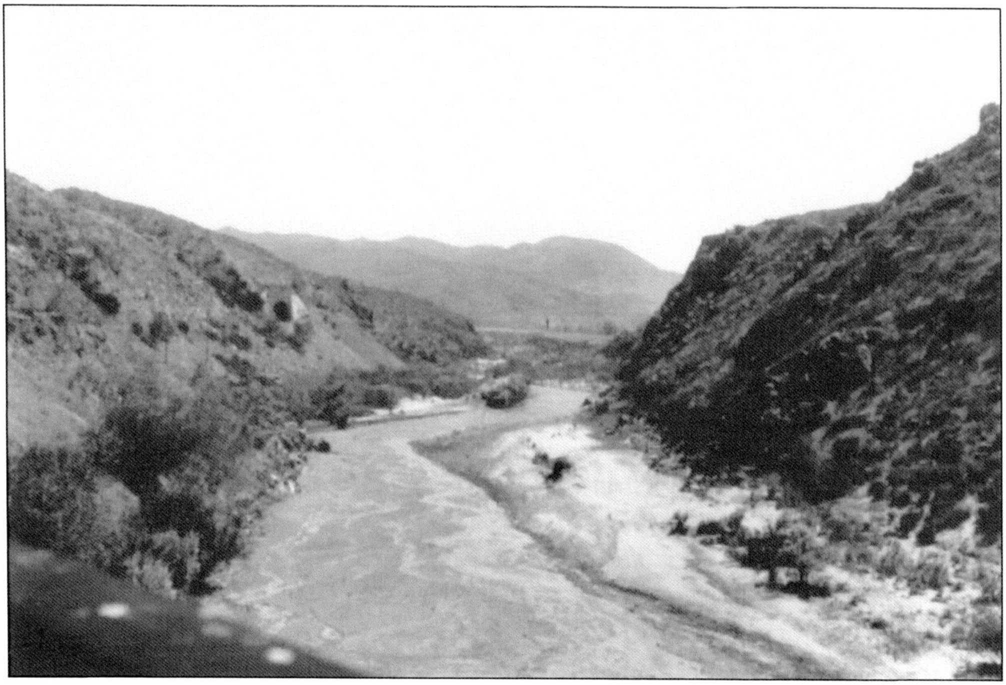

From 1900, the idea of constructing a dam on the Tijuana River, or Rio Tijuana, had intrigued San Diego and Mexican officials. The Tijuana, with its headwaters near Mount Tecate, dropped into Mexico for almost 20 miles before reentering the United States at the town of San Ysidro. Topographically, the best site was near the Marron location. However, the complicated nature of international agreements stalled the project. After years of debate, questions over logistics and funding rendered the idea problematic and squelched the project. These photographs were taken in July 1920 from the San Diego & Arizona Railroad bridge near the confluence of the Mananuco River. (Both, SDCL.)

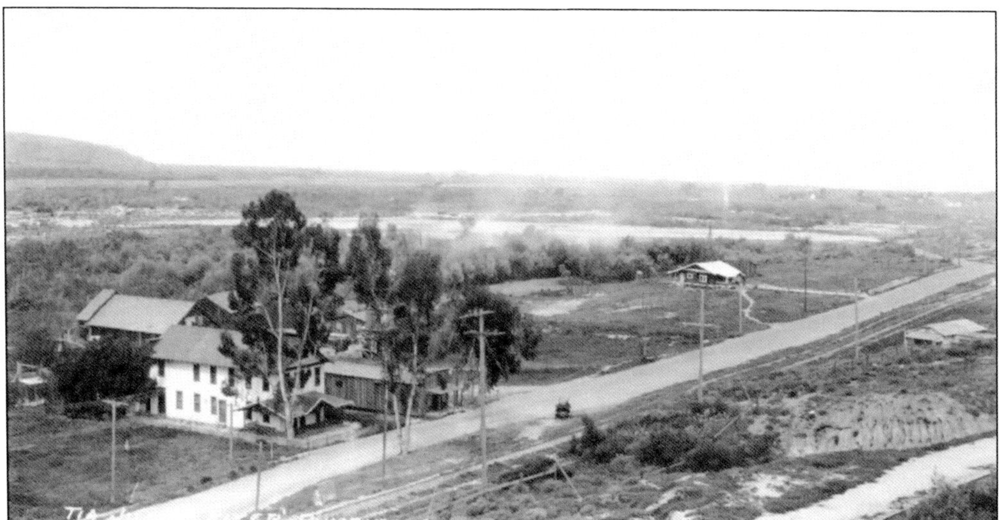

Three

THE MEN WHO BUILT THE DAMS

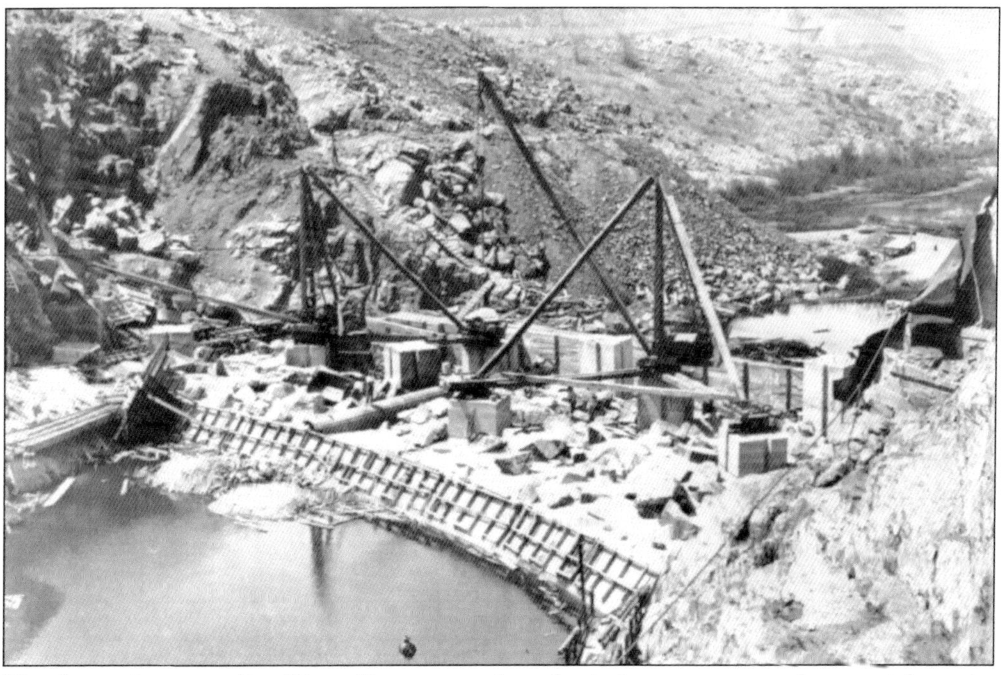

The dams of western San Diego County stand as physical monuments and memorials to the efforts of the private and civic individuals who had the foresight to attempt the ventures, the engineers who designed and supervised the construction, and particularly the men who toiled to build them. (SDCL.)

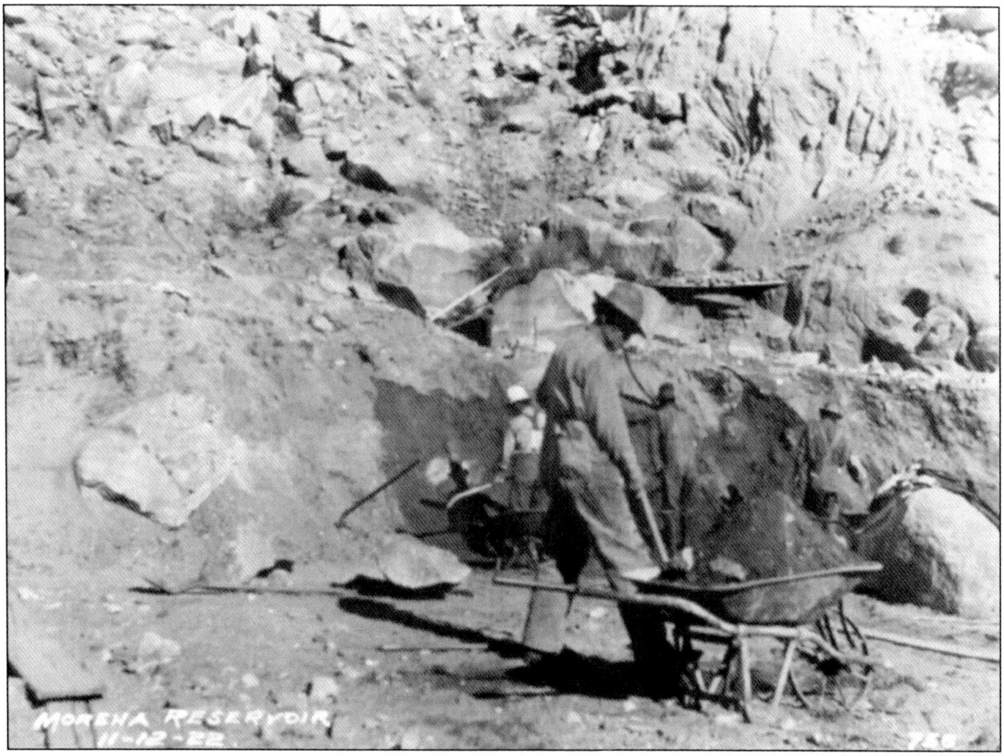

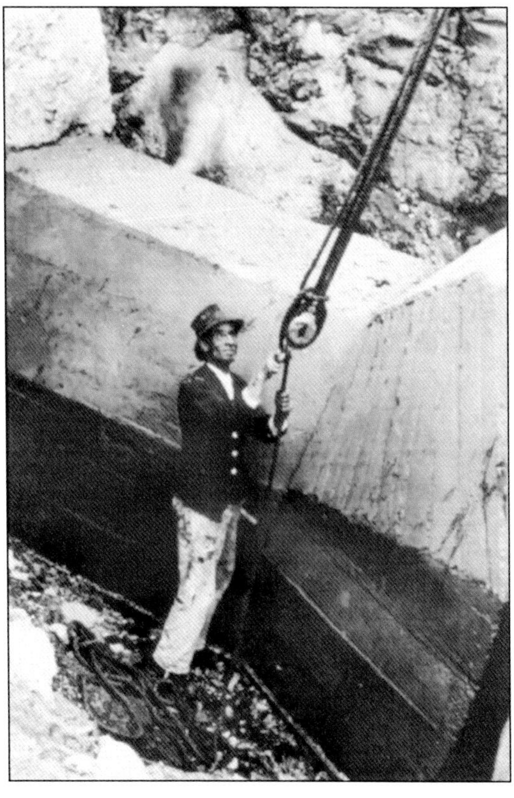

In the early era of dam construction, horses and handwork dominated a job site. The day laborers became known as "mudjacks," and concern for their safety at the construction site was largely not an issue. Unskilled workers labored in hazardous situations without protective gear. The uniform of the day was everyday clothing—leather shoes, work pants, a cotton shirt, a jacket in the winter, and a soft hat to fend off the sun, serve as a sweat band, or just keep a head warm. The fact that there were not more serious injuries is a testament—or mark of luck—to the workingman of the day. It was dangerous work at subsistence pay; laborers earned $2.50 per day and specialized workers $1 per hour. (Both, SDCL.)

Dam construction offered an accident-rich environment where men moved huge boulders, distributed tons of rock and dirt fill, worked in pits and tunnels, and trenched pipelines and dug tunnels without supports. Quarry blasting was unregulated and dangerous. Daily progress records, called 10-day reports, listed any injuries that occurred on the job, with scratches, bruises, and the occasional broken bone the most common complaints. At right, workers are placing the first pour of concrete in the foundation of a grout curtain wall, a barrier that protects the foundation from seepage. Below, men are setting the first of the massive rocks placed without mortar in the cyclopean masonry between the curtain wall and the masonry of the old dam. Both photographs were taken during the reconstruction of Lower Otay in 1918. (Both, SDCL.)

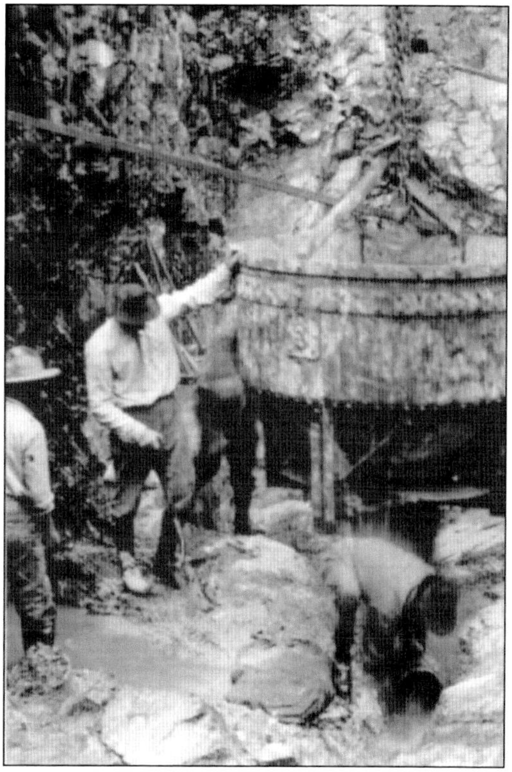

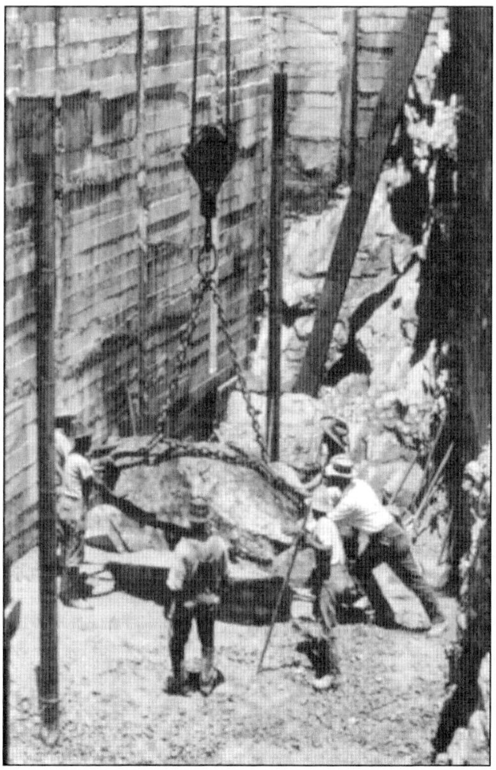

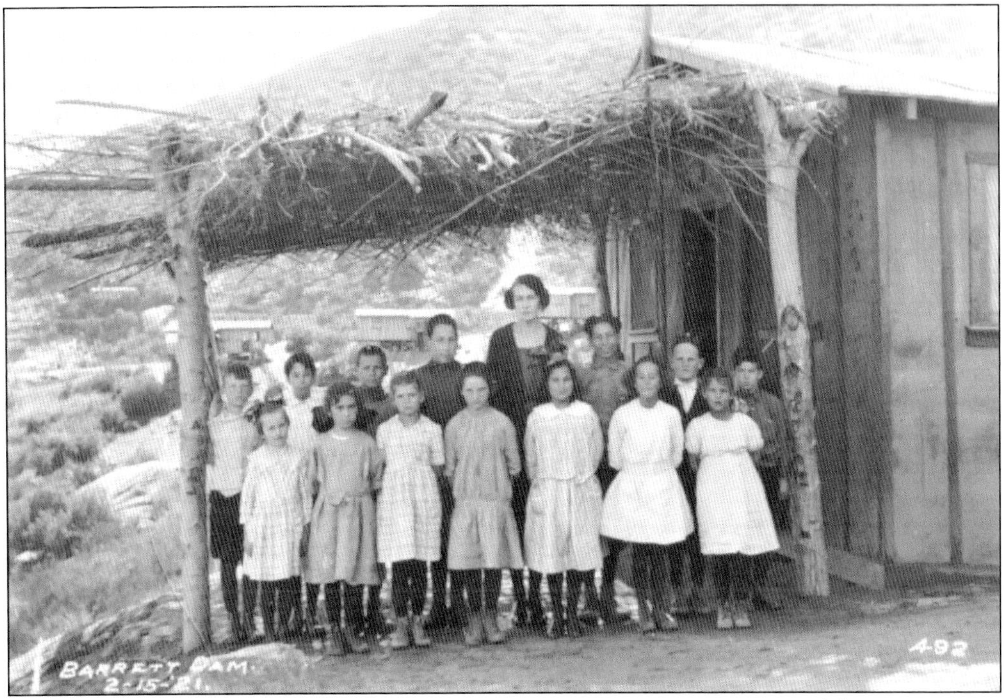

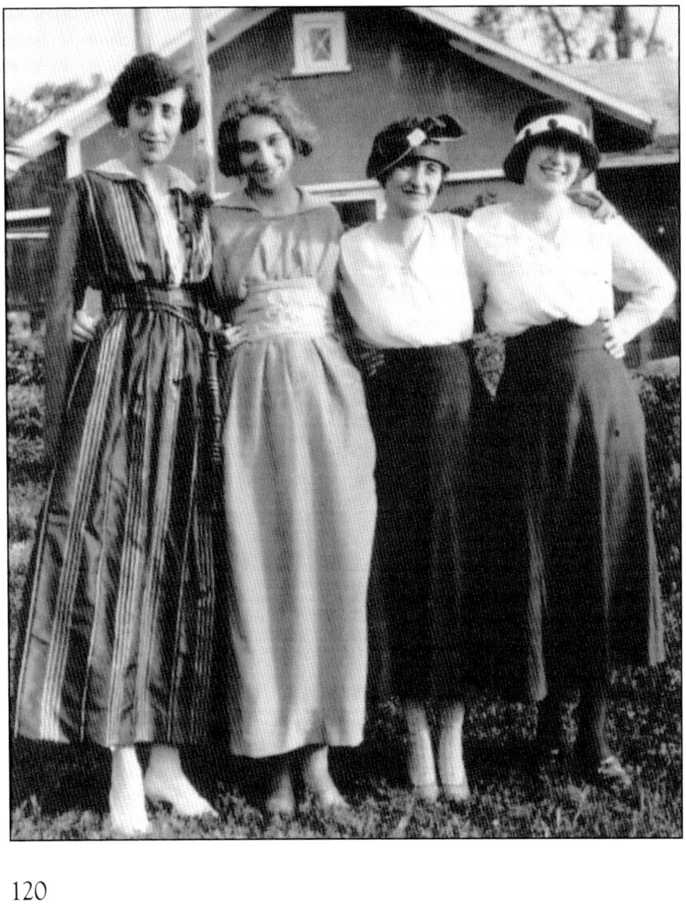

Because of the remote nature of the early dam sites, workers lived in construction camps near the job site and were afforded living accommodations, meals, and a subsistence wage. The list of hired positions included cooks, teamsters, laborers, and even a "flunky." Some camps had schools for the workers' children. Above, Myrtle Duck taught her students at the Corte Madera School at Barrett Dam construction camp in 1921. Most camps had recreational spaces, a post office, and a commissary for the laborers. A good cook was prized at the camps for both the wellbeing and morale of the workers, as was periodic entertainment. At left, this women's quartet performed for the men at Lower Otay in 1918. (Both, SDCL.)

Four

SPECTATORS, GUESTS, AND VISITORS

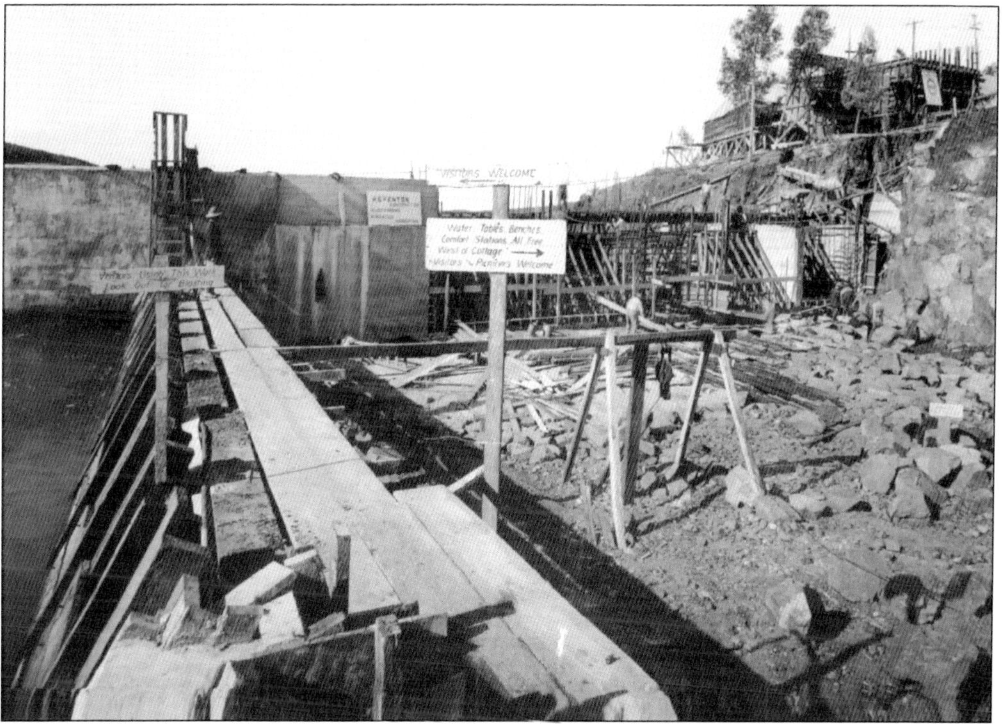

As this 1880s view at Sweetwater Dam shows, dam construction was a public event. The contractors encouraged and welcomed visitors to the construction sites. For the general population, the projects were social events. For dignitaries and politicians, they were an opportunity to be seen at an important community enterprise. For engineers, it was a chance to scrutinize and kibitz with fellow professionals. (SDCL.)

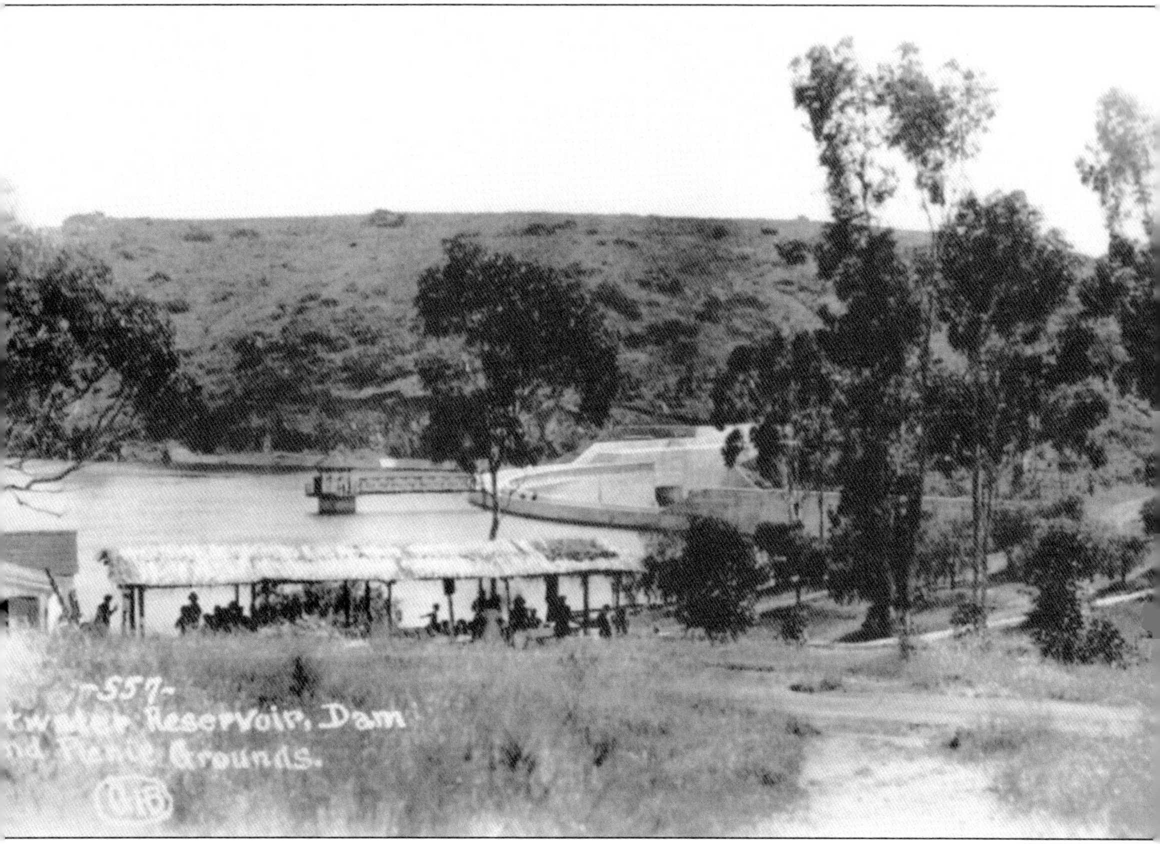

On a city project, the common council required the chief engineer to submit 10-day reports as a general accounting of the work during that period. These reports included lists of visitors, official guests, and "hydraulic engineer's guests." A 1918 report from Lower Otay listed visitors including engineers from Los Angeles and the Hodges project; city chief inspector W.H. Judy; Melville Klauber, president of the chamber of commerce; A.P. Johnson of the chamber's Water Committee; several attorneys; and other engineers and contractors. The following month, leaders of the YMCA, a group of local bankers, and members of the common council attended with the city attorney, Terence Cosgrove. Families were welcome. Here, members of the San Diego Merchants Association, the chamber of commerce, and the Kiwanis and Rotary Clubs are pictured at Barrett Dam in 1918. (Courtesy of Bonita Museum and Cultural Center.)

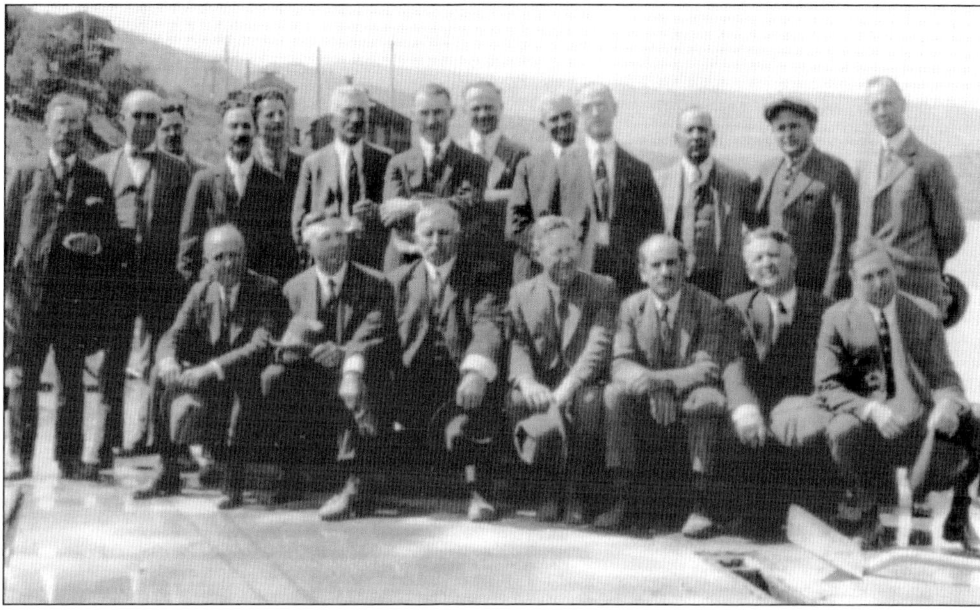

City officials generally viewed the projects as spectacles to show the public the fruits of the bonds they approved. Although both friends and foes considered Hiram Savage reserved and taciturn, he welcomed and freely posed with visitors at the construction site on the city projects he supervised between 1917 and 1934. The local newspapers offered a running log on the progress of the construction and avidly recorded the comings and goings of guests and dignitaries on the social pages. Visiting a construction site became common for civic organizations. Above, the common council joins Savage at the Otay reconstruction in 1918. Below, Savage (center) entertains (from left to right) Melville Klauber, president of the chamber of commerce; his son Hugo Klauber; engineer A.P. Johnson; and C.H. Reynolds, vice president of the Colorado Water Commission. (Both, SDCWHA.)

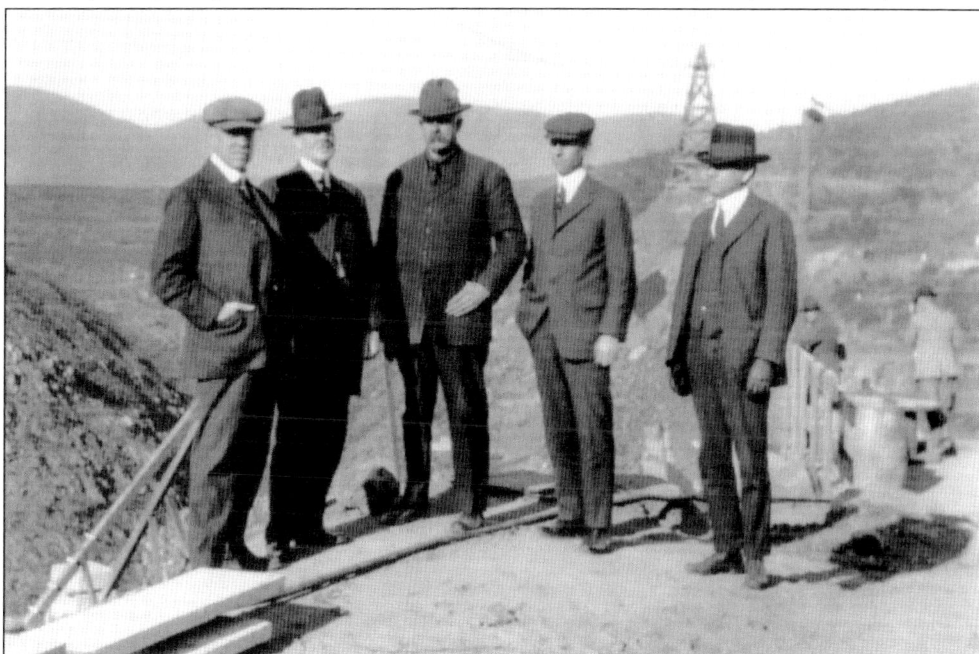

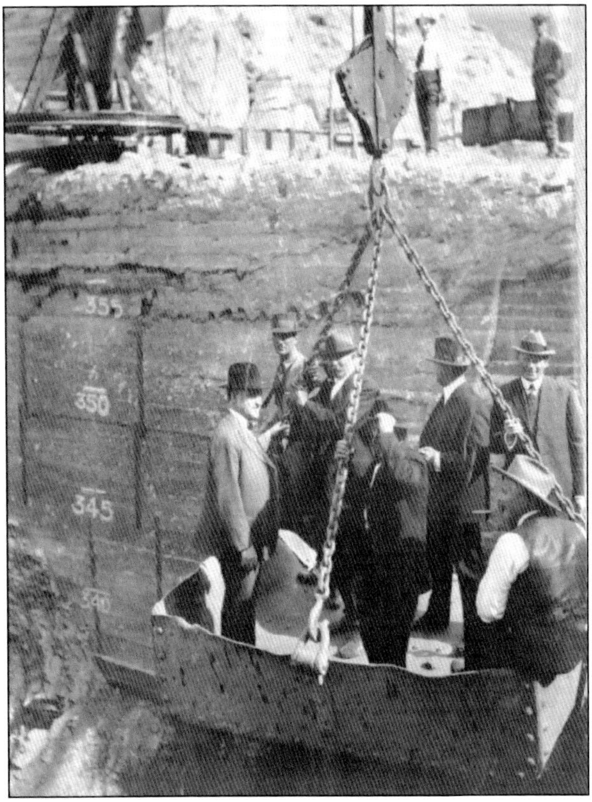

Savage appeared willing to place his guests in compromising positions just to get an eye-popping photograph and make their visit a memorable occasion. At Lower Otay in 1918, he obligingly placed his guests in the bucket derricks used to hoist rocks. At the time, insurance and safety issues took a back seat to an exciting outing to the construction site. (SDCL.)

City officials arrived to gather first-hand information, invited dignitaries reported on the status of the job, and civic groups often toured the job site to see their tax dollars at work. Here, supervising engineer Savage lectures members of the San Diego Merchants Association on the reconstruction work at Lower Otay in 1916. (SDCL.)

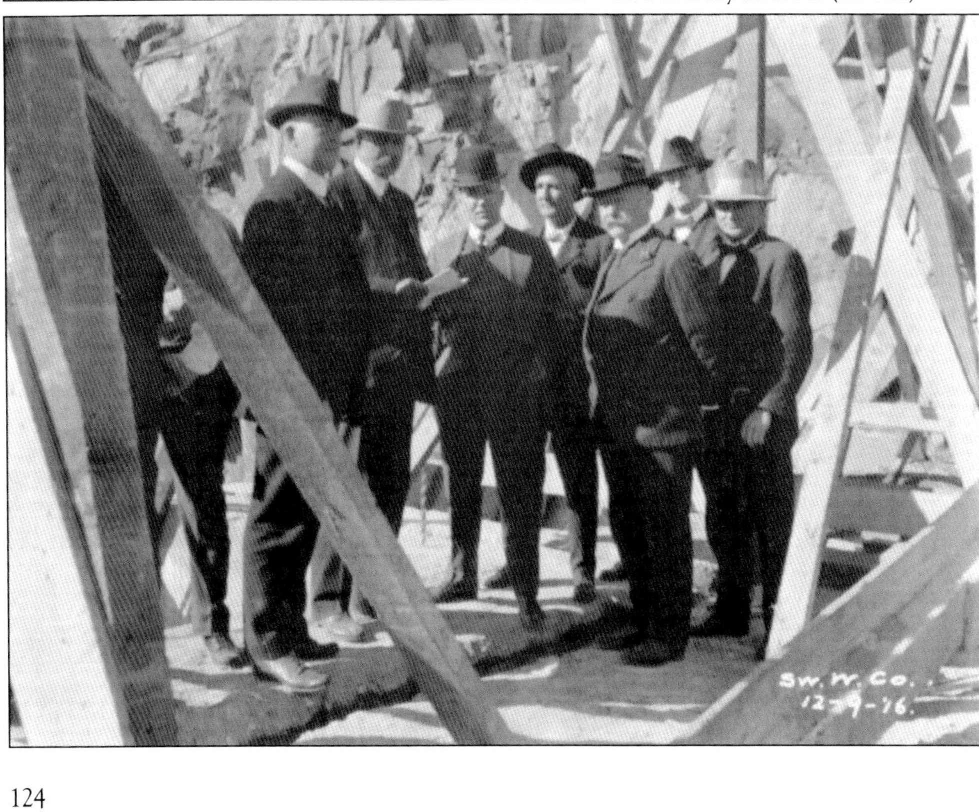

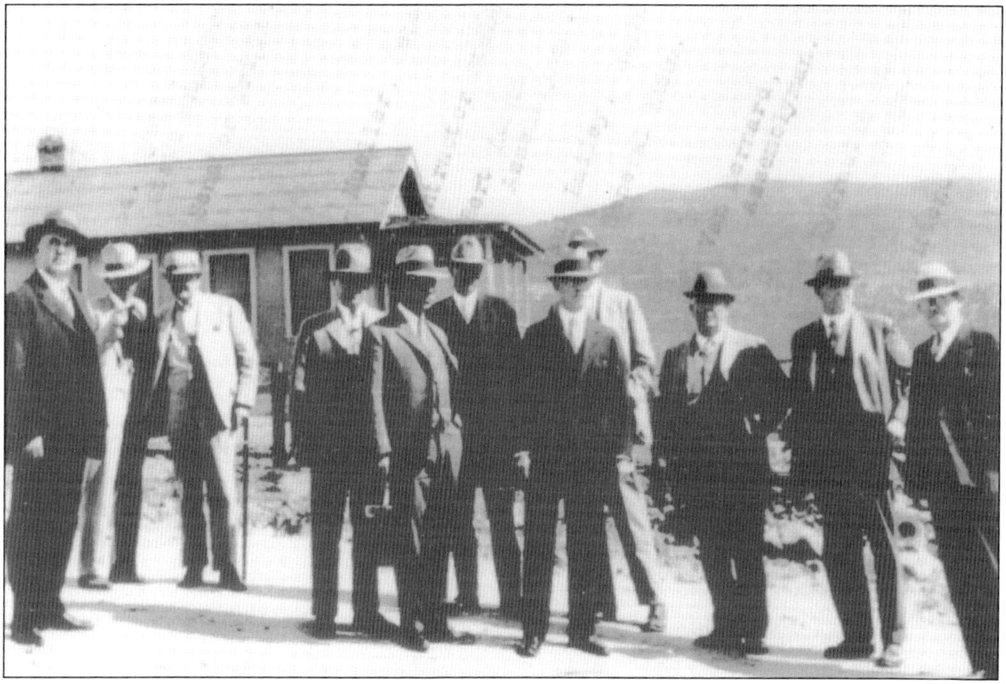

Above, a group including two US senators, three local assemblymen, and two state engineers meet with Savage to inspect Morena Dam in 1923. To accommodate these and other visiting dignitaries, every city construction site had special overnight visitors' quarters. Into the 1920s, the dams were major tourist attractions. To visit the remote Morena site, people paid $1.50 to take a stage from downtown out to the construction zone. In the photograph below taken in 1922 at Barrett Dam, Savage (far left) talks with O.E. Darnell (center) of the Security Trust Bank. E.W. Wilson of San Diego Savings Bank is in the background, and Arthur Cosgrove of Real Estate and Insurance is at far right. (Both, SDCL.)

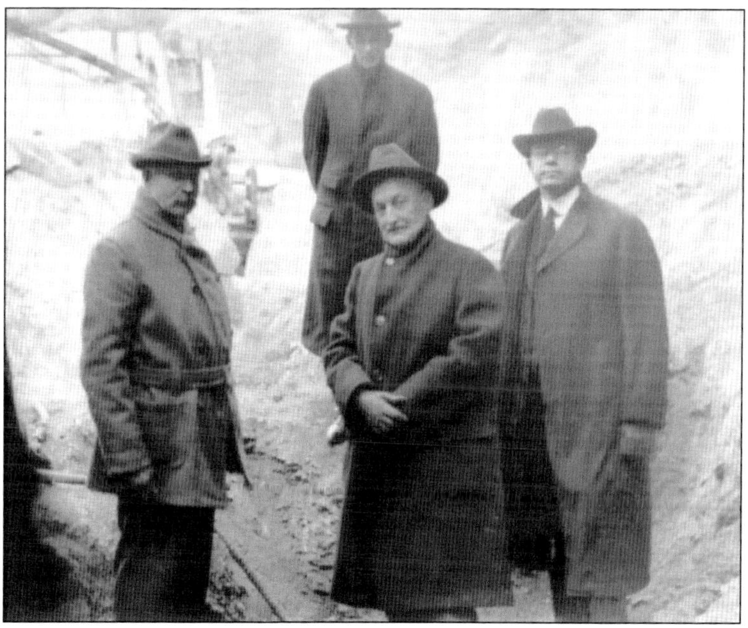

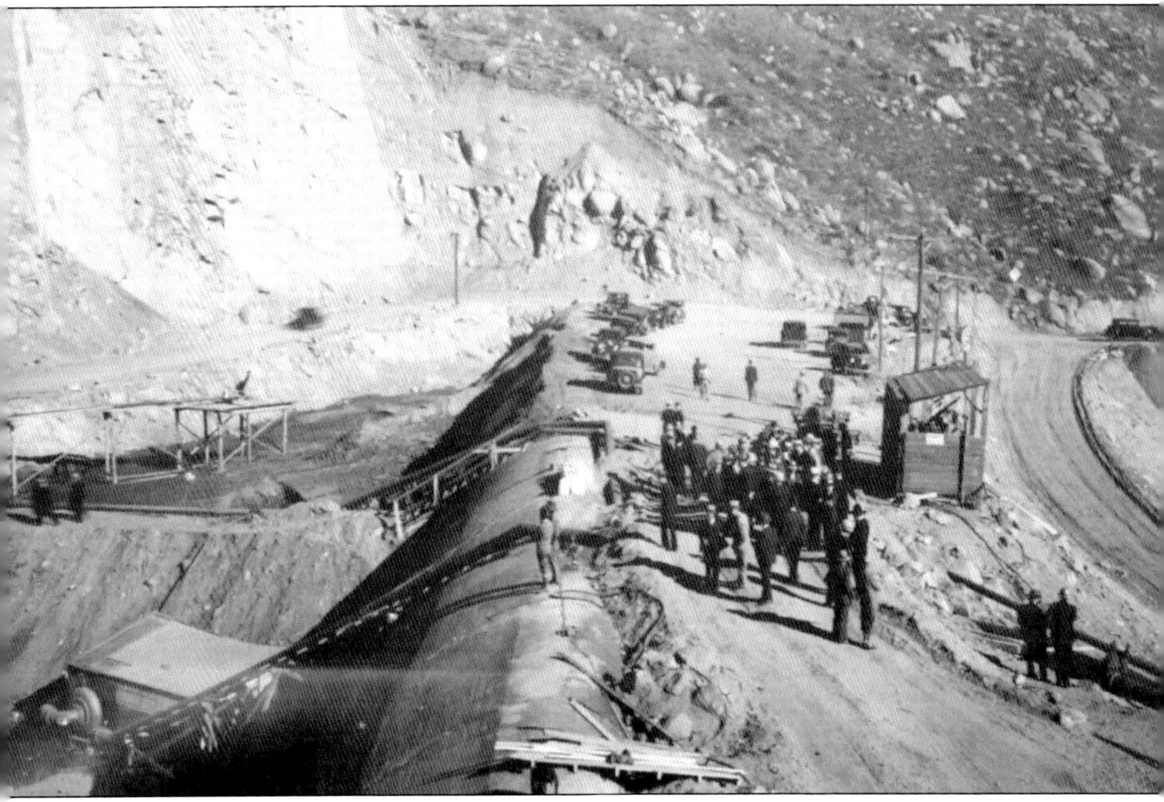

The El Capitan project was the largest dam construction venture in city history and the crown jewel of the city water system. After decades of controversy on the location of a San Diego River dam, in 1924, San Diego voters selected the El Capitan site by a three-to-one margin. The project drew intense public scrutiny. Visitors commonly undertook the long 20-mile trek from San Diego, much of it along backcountry roads and the last five miles on unimproved dirt roads, to visit El Capitan during construction in the early 1930s. On this occasion, members of the San Diego Realty Board parked their cars in the construction zone and awaited their guided tour, no doubt led by chief Hiram Savage. In this 1933 photograph, the guests are waiting on the lip of the spillway looking to the east toward the reservoir, with the dam to the south. (SDCL.)

INDEX

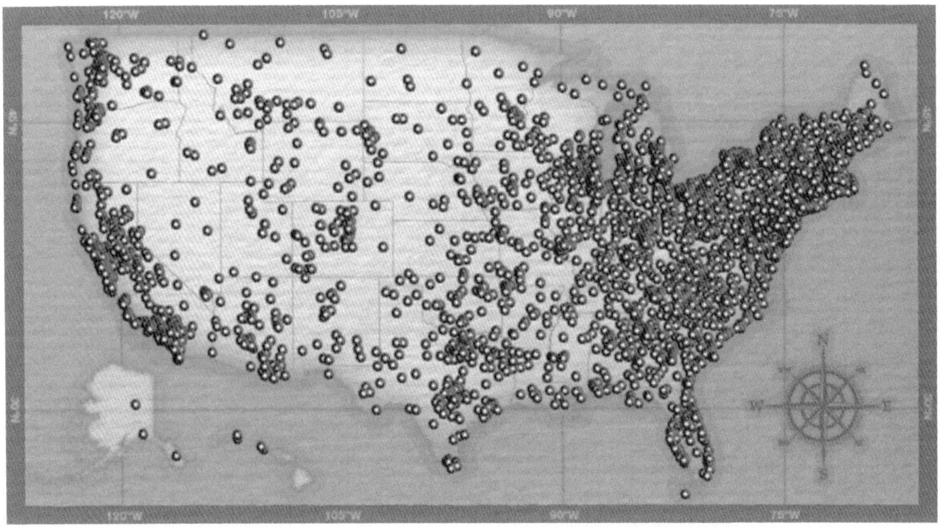